T5-AGA-411

Managing Innovation in Japan

Chihiro Watanabe

Managing Innovation in Japan

The Role Institutions Play in Helping or Hindering how Companies Develop Technology

Prof. Chihiro Watanabe
8-10-30-604 Akasaka
Minato-ku
Tokyo 107-0052
Japan
watanabe.c.aa@m.titech.ac.jp

ISBN: 978-3-540-89271-7 e-ISBN: 978-3-540-89272-4

International Institute for Applied Systems Analysis (IIASA), Laxenburg, Austria

Library of Congress Control Number: 2008941070

Springer is a part of Springer Science+Business Media

springer.com

Published by Springer-Verlag Berlin Heidelberg 2009

Cover design: WMXDesign GmbH, Heidelberg

9 8 7 6 5 4 3 2 1 0

Printed on acid-free paper

Acknowledgements

The author, Chihiro Watanabe, is grateful for the support and collaboration from the co-authors of the articles that have been published in this book. The co-authors are: B. Asgari, C. Chen, C. Griffy-Brown, J.Y. Hur, R. Kondo, K. Matsumoto, N. Ouchi and H. Wei.

Contents

Chapter 1
Introduction

This book summarizes highlights of the investigation of "*An Elucidation of the Role of Institutional Systems in Characterizing Technology Development Trajectories – A Global Comparative Analysis of Manufacturing Technology and Information Technology in the Enhancement of Business Practice*" supported by Grant-in-Aid for Scientific Research (S) by Japan's Ministry of Education, Culture, Sports, Science and Technology/Japan Society for Science Policy over the period 2002–2006.

Background and objectives of the investigation are summarized as follows:

(a) Japan ranks far below the level of the USA with respect to the development and utilization of information technology (IT) in the information society that emerged in the 1990s.
(b) This can be attributed to a vicious cycle between Japan's non-elastic institutions, insufficient utilization of the potential benefits of IT, and economic stagnation.
(c) The source of such a vicious cycle can be derived from the fundamental differences of the characterizing process of technology between manufacturing technology (MT) and IT during their diffusion processes. This investigation attempted to elucidate this mechanism.

Noteworthy findings obtained include:

(a) MT has been developed largely by the supply side and its functionality is established during the stage of its supply to the market. In contrast, IT is strongly driven by the demand side and its functionality is created through diffusion in a self-propagating way. This contrast can be clearly observed in the dramatic advancement of Japan's mobile phone industry in the late 1990s.
(b) In fact, induced by strong market learning aiming at higher functionality and accomplishing a full utilization of the benefit of network externalities, Japan's mobile phone industry has demonstrated co-evolutional development with institutional systems through the creation of functionality and the changing of characteristics in a self-propagating way during its diffusion process.
(c) This can also be attributed to co-evolution between operators and vendors amidst severe competition in their new variety of functionalities, leading to construction of Japan's unique dual co-evolutional dynamism.

C. Watanabe, *Managing Innovation in Japan: The Role Institutions Play in Helping or Hindering how Companies Develop Technology*, DOI: 10.1007/978-3-540-89272-4_1,

(d) Mobile phone driven innovation triggered reactivation of Japan's economy through its catalytic role in restructuring industry production, diffusion, consumption and innovation systems by disseminating core technologies and renovating the conventional production style.
(e) However, as a consequence of Japan's unique supply structure in a closed system, Japan's mobile phone industry failed to conform to the demand for functionality variety in the global market.
(f) In addition, innovation resulted in differentiating the domestication of the characterization process of technology between firms sticking to not invented here syndrome and those that succeeded in fusing indigenous strength and learning effects by means of the hybrid management of technology. Consequently, the innovation trajectory of firms was bi-polarized, resulting in heterogeneous Japanese firms that were previously homogenous.

Chapter 2
Formation of IT Features Through Interaction with Institutional Systems: Empirical Evidence of Unique Epidemic Behavior

Abstract While emerging information technology (IT) is hastening the paradigm shift from an industrial society to an information society and providing all nations of the world with numerous potential benefits, effective utilization of these benefits will differ greatly depending on the nation, particularly on their institutional elasticity. This can be attributed to the specific features of IT. Since IT performs its function in connection with institutional systems unlike technology in general, its specific features can be formed through dynamic interaction with an institutional system. Considering the unique features of IT formed through such dynamic interaction, this chapter focuses on an analysis of the epidemic behavior of IT and attempts to identify specific features of IT in light of interaction with institutions.

Reprinted from *Technovation*, 23 No. 3, C. Watanabe, R. Kondo, N. Ouchi and H. Wei, Formation of IT features through interaction with institutional systems – Empirical evidence of unique epidemic behavior, pages: 205–219, copyright (2003), with permission from Elsevier.

2.1 Introduction

Innovation, such as the development of new technologies, has been undoubtedly recognized as a significant driving force in sustaining economic growth. Romer [1] points out that for society as a whole, innovation, discovery and technological change offers large net gains because the new goods or processes are more efficient and more valuable than the old ones. On the other hand, as the OECD [2] claims, growth depends on building and maintaining an environment that is conductive to innovation and the application of new technologies. Actually, new technology itself represents only potential, and in order to exploit such potential, institutional change is necessary [3].

A rapid surge in IT around the world is inevitably forcing traditional societies to transform their socioeconomic structures. As the Telecommunications Council [4] noted, IT is hastening the paradigm shift from an industrial society to an information

C. Watanabe, *Managing Innovation in Japan: The Role Institutions Play in Helping or Hindering how Companies Develop Technology*, DOI: 10.1007/978-3-540-89272-4_2,

society. However, if IT is merely introduced to replace part of the workforce so as to improve productivity, as was the case with automation, the full benefits of IT will not be utilized. This is because IT not only enhances task efficiency but also permeates through an organization, or a society, to have an impact on their structure and behavior. More precisely, IT waves, most recently exemplified by growing popularity of the Internet and mobile communications are characterized by so-called "network externalities"[1] [5, 6] that construct a virtuous cycle between expanding number of users and rising value of networks, and rapidly diffuse as social infrastructure to support socio-economic activities.

The OECD [3] analyzed the potential of IT to "automate" and "informate." It observed that more relative emphasis has been given to the "automate" option and that IT has often been introduced into organizations that were shaped independently of it. Thus, if an organization can reengineer itself to shift the balance away from the "automate" option towards the "informate" option, it can become a learning institution with a new sets of skills.

Accordingly, IT differs from other technologies in that it interacts with individuals, organization, and societies (or institutions) in order to be utilized, and its features are formed dynamically through this interaction, behaving differently depending on the institutional elasticity of that it interacts with. In other words, the unique features of IT are formed during the course of interaction with institutional systems [6, 7]. Consequently, *IT's unique features can be identified in its diffusion process with respect to individuals, organizations and societies.*

Research on *the diffusion of innovation* has been undertaken in broad fields independently for long years including anthropology, sociology, education, public health, communication, marketing, and geography. Rogers [8] attempted to systematize these works in his pioneer work in "Diffusion of Innovations". He defined "diffusion" as *the process by which an innovation is communicated through certain channels over time among the members of a social system.* He also identified four main elements in the diffusion of innovations: *innovation features, communication channels, time, and social system.* All of Rogers's postulates support our hypothetical view with respect to IT features formation process that IT's unique features can be identified in its diffusion process.

This diffusion process is actually quite similar to the contagion process of an epidemic disease [9] and exhibits S-shaped growth. This process is well modeled by the *simple logistic growth function*, an epidemic function which was first introduced by Verhulst in 1845 [10]. Since the logistic growth function has proved useful in modeling a wide range of innovation process, a number of studies applied this function in analyzing the diffusion process of innovations [9], Mansfield [11, 12], Metcalfe [13], Norris and Vaizey [14].

While the simple logistic growth function treats the carrying capacity of a human system[2] fixed, this capacity is actually subject to change [15]. Among varieties of innovations, certain innovations alter their carrying capacity in the process of

[1] The value to a consumer of a product increases as the number of compatible users increases [20].

[2] Upper limit of the level of diffusion, see Sect. 2.2.2.

their diffusion which stimulates increase in the number of potential customers [16]. This increase, in turn, incorporates new features to the innovations. This is similar to an ecosystem in which species can sometimes alter and expand their niche [10]. Mayer [10] extended the analysis of logistic functions to cases where dual processes operate by referring an example when cars first replaced the population of horses but then took on a further growth trajectory of their own. He stressed that "if the carrying capacity of a system changes during a period of logistic growth, a second period of logistic growth with a different carrying capacity can superimpose on the first growth pulse." This is quite similar to the diffusion process of the innovations discussed above. Aiming at exhibiting this diffusion process that contains complex growth processes not well modeled by the single logistic, Mayer postulated *bi-logistic growth*.

In addition to the above diffusion processes exhibited by a single logistic growth and bi-logistic growth, in particular innovations, correlation of the interaction between innovations and institutions display systematic change in their process of the growth and maturity. This is typically the case of the diffusion process of IT in which *network externality* [17] functions to alter the correlation of the interaction which creates new features of the innovation, IT. In this case, the rate of adoption increases, usually exponentially until physical or other limits slow the adoption. Adoption is a kind of "social epidemic." Schelling [18] portrays an array of logistically developing and diffusing social mechanisms stimulated by these efforts. Meyer and Ausbel [19] introduced an extension of the widely used logistic model of growth by allowing it for a sigmoidally increasing carrying capacity. They stressed that "evidently, new technologies affect how resources are consumed, and thus if carrying capacity depends on the availability of that resource, the value of the carrying capacity would change." This explains, particularly, unique diffusion process of IT which diffuses by altering carrying capacity or creating a new carrying capacity in the process of its diffusion. Aiming at exhibiting this diffusion behavior, Mayer and Ausbel proposed *logistic growth within a dynamic carrying capacity*.

Provided that the unique features of IT are formed during the course of interaction with institutional systems and that these features can be identified in its diffusion process, we can expect to identify features formation process of IT and its specific features in light of interaction with institutions by analyzing its diffusion trajectory using logistic growth within a dynamic carrying capacity and by comparing it with diffusion processes of technology in general. Furthermore, given the significance of network externality and its contribution to enhance IT's carrying capacity, identification of the mechanism of IT's contribution to increasing returns to scale can be expected.

In light of the increasing importance of institutional elasticity in exploiting the full potential of IT, this paper attempts to derive specific IT features by focusing on the unique diffusion process, in other words epidemic behavior of IT.

Section 2.2 conducts comparative analysis of epidemic behavior between IT and other technologies. Section 2.3 extracts implications with respect to features formation process of IT and its specific features. Section 2.4 briefly summarizes the key findings of the analysis, presents conclusions and discusses implications for effective utilization of the potential benefits of IT.

2.2 Features of IT with Respect to Institutions

2.2.1 Formation Process of Specific Features of Technology

As repeatedly emphasized in numerous studies, IT is functioning as a driving force to transform the existing socioeconomic structure by permeating through people's daily life, organizational activities, and society as a whole, hastening the paradigm shift from an industrial society to an information society [4, 6, 7, 20].

Table 2.1 compares features of the core technologies in the 1980s and in the 1990s. During the 1980s, developing excellent manufacturing technology was a key for firms to be successful in an industrial society. Manufacturing technology has been developed by a supply side to provide end-users with products or has been introduced to factories to replace part of the workforce for improving productivity. Like other technologies, features of manufacturing technology are established or programmed at birth and once it leaves a supply side, it would not change its behavior substantially during its dissemination. In this case, individual firms are responsible for forming features of technology.

With the remarkable development of information technology, IT, especially increased electronic connectivity in the 1990s, socio-economic activities have been more relying on IT infrastructures. The worldwide Internet population has been increasing[3] and the Internet has made it easier and cheaper for all businesses to transact business and exchange information, leading to an expanding e-commerce market [6].

Contrary to manufacturing technology, suppliers of IT are more concerned about compatibility. This is because IT products are often utilized as communication tool. If an electronic file processed by NEC's personal computer is not compatible with that of Toshiba's, how valuable these computers are? If a subscriber to a certain mobile communications service career cannot make a call to a subscriber of another career, people should lose an incentive to purchase cellular telephones, or try to subscribe to a career that boasts dominant number of subscribers. On the other hand, any home appliances such as refrigerators or TV sets can be purchased without

Table 2.1 Comparison of features between manufacturing technology and IT

	1980s	1990s
Paradigm	Industrial society	Information society
Core technology	Manufacturing technology	IT
Key features	*Given, provided by suppliers*	*To be formed during the course of interaction with institutions*
Actors responsible for formation of features	Individual firms/organizations	Institutions as a whole

[3] According to Nua Internet Surveys, there were approximately 407.1 million internet users world wide as of November 2000.

being bothered by what other people possess. In this context, IT products are subject to phenomena so-called network externalities. With computers and telephones, for example, the more people use compatible systems or the more people are on a network, the more valuable the system or the network become, thus attracting more potential users [5].

In short, IT strongly possesses a self-multiplicative feature that closely interacts with individuals, organizations, and society in broad, institutions, during the course of its diffusion and behaves differently depending on institutions of that it interacts with. These observations suggest that features of IT are formed dynamically during the course of interaction with institutions and whether the potential benefits of IT can be exploited greatly depends on institutions.

This formation process of IT features is actually quite similar to the contagion process of an epidemic disease stimulated by this similarity and based on the above hypothetical view that unique features of IT are formed during the course of its dissemination process, an attempt to derive specific features of IT with respect to the interaction with institutions by examining the unique epidemic behavior of IT is conducted.

2.2.2 *Analysis of Epidemic Behavior*

2.2.2.1 Taxonomy of Epidemic Function

Following three functions as introduced in Sect. 2.1 were used for comparative analysis of epidemic behaviors between IT and other technologies:

Simple logistic growth function: $f(t) = \frac{K}{1+a\exp(-bt)}$ *where a and b: coefficients; and t: time trend.*

An epidemic function is used for analyzing the diffusion and maturity of innovative goods. The epidemic function enumerates the contagion process of an epidemic, and this model provides an analogy of the diffusion and maturity trajectory through the contagion process of innovative goods similar to a medical epidemic. The epidemic function incorporates a negative feedback in an exponential function as follows:

$$\frac{\mathrm{d}f(t)}{\mathrm{d}t} = bf(t)\left(1 - \frac{f(t)}{K}\right), \tag{2.1}$$

where K indicates the upper limit of $f(t)$: carrying capacity.

$\left(1 - \frac{f(t)}{K}\right)$ depicts a negative feed back and this approaches 1 and 0 when $f(t) \to K$ and $f(t) \to K$, respectively. Therefore, the growth rate (the left hand side of (2.1) increases logistically at the initial stage and stagnates to 0 as $f(t)$ approaches to K, drawing an S-shaped curve as illustrated in Fig. 2.1.

The following equation can be obtained by integrating (2.1):

$$f(t) = \frac{K}{1 + a\ \exp(-bt)}. \tag{2.2}$$

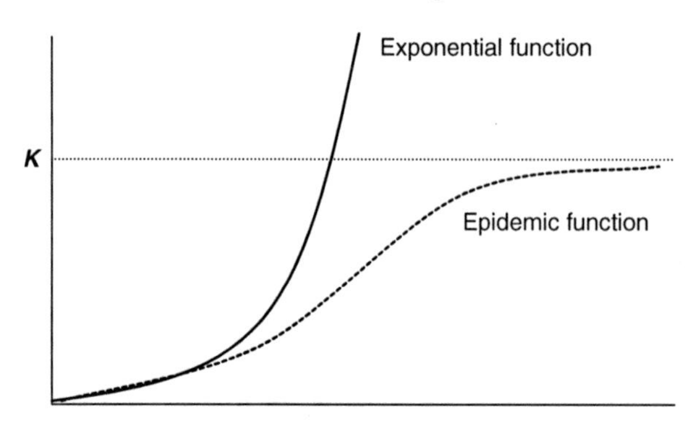

Fig. 2.1 Comparison between exponential function and epidemic function

Bi-logistic growth function: $f(t) = f_1(t) + f_2(t) = \frac{K_1}{1+a_1\ \exp(-b_1t)} + \frac{K_2}{1+a_2\ \exp(-b_2t)}$.
Bi-logistic growth function combines two phases of simple logistic growth function in the following function [10]:

$$f(t) = f_1(t) + f_2(t) = \frac{K_1}{1+a_1\ \exp(-b_1t)} + \frac{K_2}{1+a_2\ \exp(-b_2t)}. \tag{2.3}$$

Logistic growth function within a dynamic carrying capacity:

$$f(t) = \frac{K_K}{1+a\ \exp(-bt) + \frac{b \cdot a_k}{b-b_K}\exp(-b_Kt)}.$$

The epidemic function expressed by (2.1) assumes that the level of carrying capacity (K) is constant through the dissemination process of innovation. However, as reviewed in Sect. 2.1, in particular innovations, correlation of the interaction between innovation and institutions display systematic change in their process of the growth and maturity leading to creating new carrying capacity in the process of its diffusion. In these innovations, the level of carrying capacity will be enhanced as their diffusion proceed, and carrying capacity K in (2.1) should be treated as a following function:

$$\frac{\mathrm{d}f(t)}{\mathrm{d}t} = bf(t)\left(1 - \frac{f(t)}{K(t)}\right), \tag{2.4}$$

where $K(t)$ is also an epidemic function enumerated by (2.5).

$$K(t) = \frac{K_K}{1+a_K\exp(-b_Kt)}, \tag{2.5}$$

where K_k indicates the ultimate upper limit.

The solution of a differential (2.4) under the condition (2.5) can be obtained as an (2.6).[4]

$$f(t) = \frac{K_K}{1+a\ \exp(-bt) + \frac{b \cdot a_k}{b-b_K}\exp(-b_Kt)}. \tag{2.6}$$

[4] See Appendix 1 for details of mathematical development.

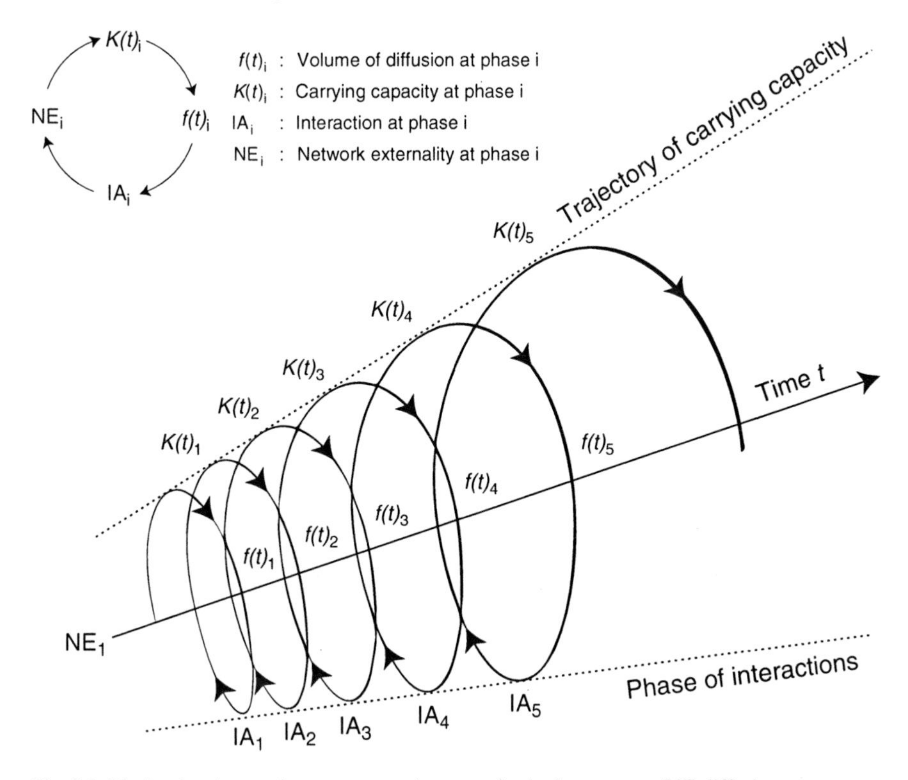

Fig. 2.2 Mechanism in creating a new carrying capacity in the process of IT diffusion

A dynamic carrying capacity $K(t)$ can be expressed by (2.7) by transforming (2.4).

$$K(t) = f(t)\left(\frac{1}{1-\left(\mathrm{d}f(t)/\mathrm{d}t\right)/bf(t)}\right). \tag{2.7}$$

Equation (2.7) demonstrates that $K(t)$ increases together with the increase of $f(t)$ as time goes by. This implies that (2.6) exhibits logistic growth within a dynamic carrying capacity as is displays such systematic change as illustrated in Fig. 2.2 number of customers (volume of diffusion) increases as time passes, which indicates interactions with institutions leading to increasing potential customers (carrying capacity) by increased value and function stimulated by network externality. Thus, IT's specific features are formed in this process.

2.2.2.2 Comparative Analysis of Epidemic Behavior

In order to verify the difference in diffusion process between IT and other technologies, diffusion patterns of (1) refrigerators, (2) color TV sets, and (3) cellular telephones were analyzed by applying the above mentioned three models. In the analysis, refrigerators and color TV sets were chosen because they are regarded as representative products of manufacturing technology, while cellular telephones represent one of the most popular IT products.

As indicators to measure diffusion patterns in Japan, annual shipments (1966–1999), annual domestic shipments (1966–2000), and quarterly domestic production volumes deducting imports and exports (1993–2000) were used for refrigerators, color TV sets, and cellular telephones, respectively.[5]

Refrigerators

Among three models, simple logistic growth function was most fitting for the diffusion pattern of refrigerators in that statistical indicators such as t-value and AIC (Akaike's Information Criterion) were most significant compared with those of other models, and adj.R^2 and DW were also relatively significant. As for the application of logistic growth function within a dynamic carrying capacity, a_k was small enough to lead the carrying capacity to almost fixed and the resultant trajectory traced a similar curve as simple logistic growth function did.

Figure 2.3 illustrates the trends in diffusion process of refrigerators in Japan from the year 1966 to 1999 with simple logistic growth function (figures in parentheses indicate t-value).

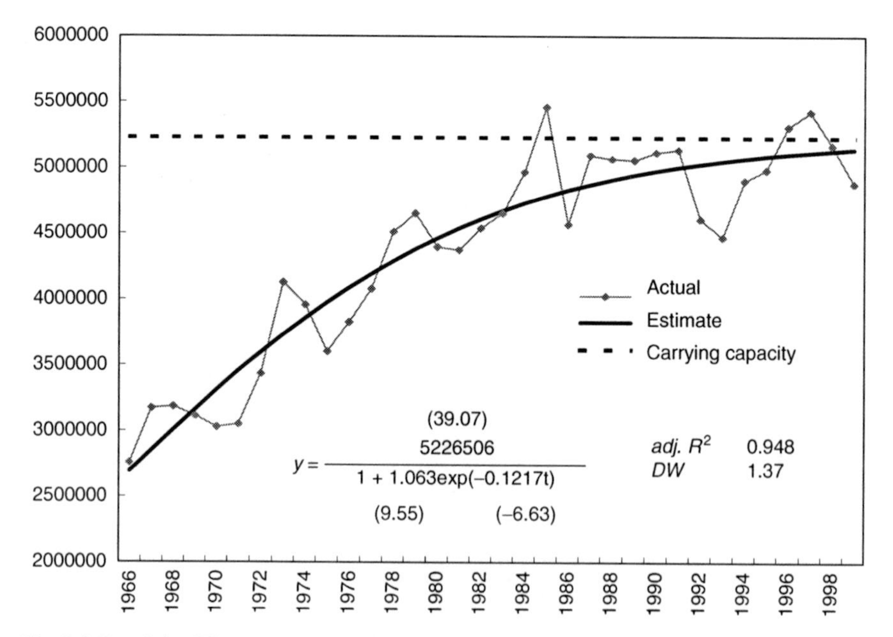

Fig. 2.3 Trends in diffusion process of refrigerators in Japan (1966–1999)
Source: Report on machinery statistics, MITI (annual issues)

[5] See Appendix 2 for data construction and sources.

Color TV sets

The diffusion process of color TV sets most fitted bi-logistic growth function for which all the statistical indicators showed significant values. Figure 2.4 depicts the transition of annual domestic shipment of color TV sets and related events from the year 1966 to 2000. In Japan, color TV broadcasting service was started in 1960. Since then, viewers gradually switched their TV sets from monochrome to color triggered by events such as World Exposition of Osaka in 1970. In 1973, color TV broadcasting became available in all TV programs, and viewers has come to be more concerned with the contents of TV programs, not color TV sets themselves.

Figure 2.5 illustrates the trends in diffusion process of color TV sets in Japan from the year 1966 to 2000 with bi-logistic growth function.

Cellular Telephones[6]

Figure 2.6 depicts the transition of quarterly production volume of cellular telephones and related events from the year 1993 to 2000. Though cellular telephones have relatively young history compared with that of refrigerators and color TV sets, continuous development of smaller and lighter handsets with a variety of functions has made their diffusion process rather complicated. One of the breakthroughs was NTT DoCoMo's introduction of i-mode service in February 1999 that enabled users to access to the Internet from their handsets. Since then, this kind of mobile

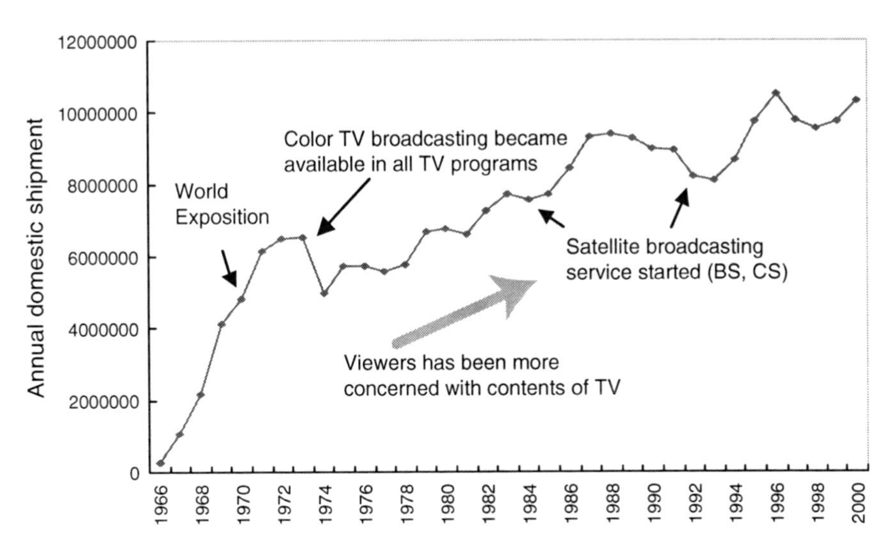

Fig. 2.4 Transition of annual domestic shipment of color TV sets and related events
Source: Japan Electronics and Information Technology Industries Association, Japan

[6] Cellular telephones include PHS (personal handy-phone systems) and automobile phones as well as cell phones.

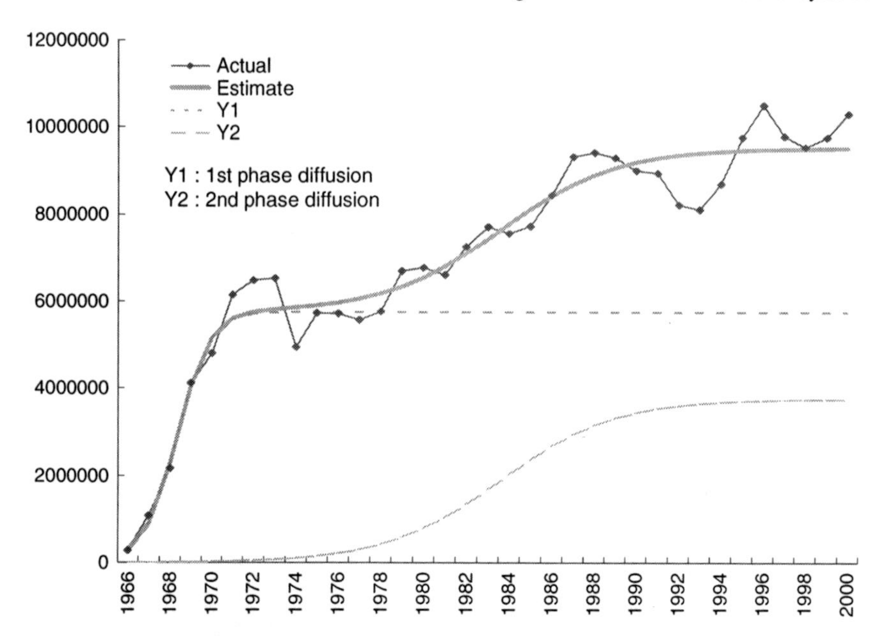

Fig. 2.5 Trends in diffusion process of color TV sets in Japan (1996–2000)
Source: Japan Electronics and Information Technology Industries Association, Japan

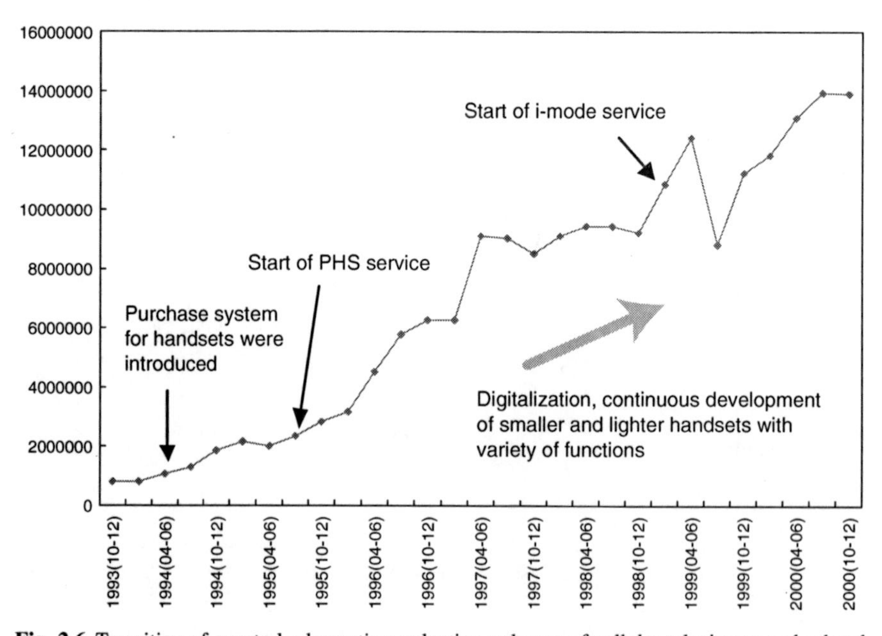

Fig. 2.6 Transition of quarterly domestic production volumes of cellular telephones and related events
Source: Current survey of production, METI (annual issues); trade statistics, MOF (annual issues)

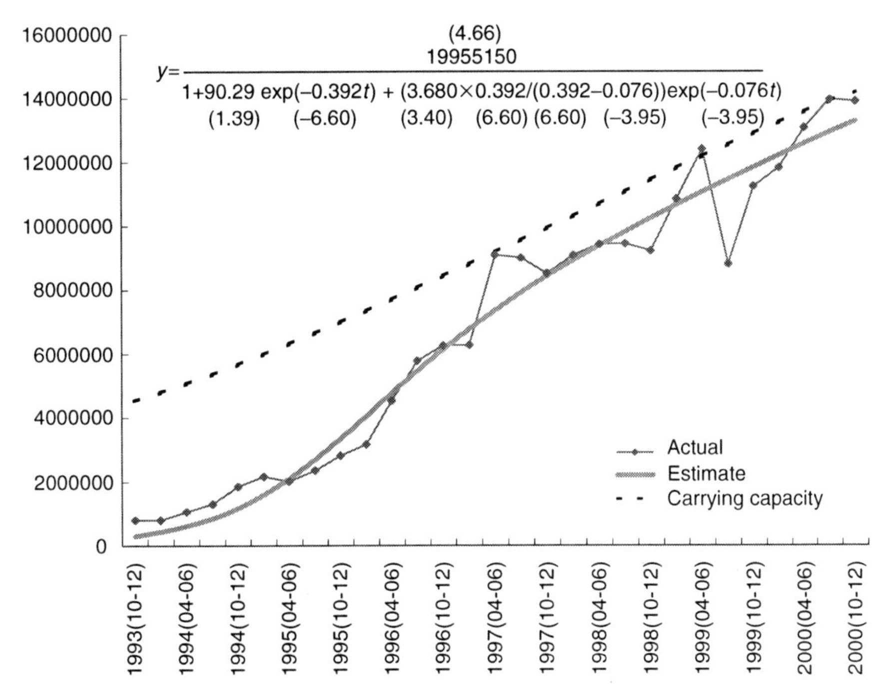

Fig. 2.7 Trends in diffusion process of cellular telephones in Japan (1993–2000)
Source: Current survey of production, METI (annual issues); trade statistics, MOF (annual issues)

Internet access service has been dramatically expanding and the number of subscribers reached about 31.4 million as of February 2001.[7] Java-compatible handsets have now been on sale since January 2001, which is expected to induce further increase in carrying capacity.

With these features, the diffusion process of cellular telephones was most fitting to logistic growth function within a dynamic carrying capacity, which showed the least AIC value among other models. Although adj.R^2 and DW were relatively significant for bi-logistic growth function, AIC is more reliable since the data was analyzed by non-linear regression. Figure 2.7 illustrates the trends in diffusion process of cellular telephones in Japan from the year 1993 to 2000 with logistic growth function within a dynamic carrying capacity.

2.2.2.3 Interpretations

Table 2.2 compares fittability of three epidemic functions: (1) Simple logistic growth function, (2) bi-logistic growth function, and (3) Logistic growth function within a dynamic carrying capacity for the diffusion process of three innovative goods: refrigerators, color TV sets, and cellular telephones.

[7] http://www.tca.or.jp/

Table 2.2 Comparison of the fittability of three epidemic functions for the diffusion process of three innovative goods

Refrigerators								
(1)	K	a	b				$adj.R^2$	AIC
	5,226,506	1.063	0.1217				0.948	8.523×10^{10}
	(39.07)	(9.55)	(6.63)					
(2)	K_1	a_1	b_1	K_2	a_2	b_2	$adj.R^2$	AIC
	3,399,421	0.2436	0.1101	1,677,219	22.17	0.2842	0.926	9.183×10^{10}
	(21.33)	(2.22)	(0.50)	(21.33)	(0.79)	(3.89)		
(3)	K_K	a	b	a_k	b_k		$adj.R^2$	AIC
	5,226,506	0.4637	0.1217	4.141×10^{-6}	0.1217		0.949	8.706×10^{10}
	(38.97)	(4.46)	(6.62)	(1.50)	(6.62)			
Color TV sets								
(1)	K	a	b				$adj.R^2$	AIC
	9,867,687	3.141	0.1385				0.923	1.018×10^{12}
	(18.61)	(5.60)	(5.43)					
(2)	K_1	a_1	b_1	K_2	a_2	b_2	$adj.R^2$	AIC
	5,758,095	71.90	1.281	3,759,613	973.0	0.3701	0.966	3.896×10^{11}
	(19.99)	(1.03)	(4.45)	(8.49)	(0.40)	(2.75)		
(3)	K_K	a	b	a_k	b_k		$adj.R^2$	AIC
	9,867,687	0.2109	0.1386	0.002653	0.1385		0.955	6.287×10^{11}
	(18.61)	(0.39)	(5.43)	(0.83)	(5.43)			
Cellular telephones								
(1)	K	a	b				$adj.R^2$	AIC
	13,328,320	18.07	0.203				0.980	1.014×10^{12}
	(16.59)	(3.81)	(7.54)					
(2)	K_1	a_1	b_1	K_2	a_2	b_2	$adj.R^2$	AIC
	9,944,280	32.40	0.308	12,016,010	10,538	0.299	0.993	7.102×10^{11}
	(16.42)	(2.64)	(7.23)	(16.42)	(1.12)	(9.20)		
(3)	K_K	a	b	a_k	b_k		$adj.R^2$	AIC
	19,955,150	90.29	0.392	3.680	0.076		0.984	6.370×10^{11}
	(4.66)	(1.39)	(6.60)	(3.40)	(3.95)			

Figures in parentheses indicate t-value.

Looking at Table 2.2, we note the following findings with respect to the identification of epidemic behavior for respective innovative goods:

(a) AIC suggests that simple logistic growth function for refrigerators, bi-logistic growth function for color TV sets, and logistic growth function within a dynamic carrying capacity for cellular telephones demonstrate most fittable functions, respectively.

(b) Refrigerators have single function and more than half a century history since they have penetrated in a market. Their diffusion volume has almost saturated to 5.2 ± 0.5 million in the last 15 years. Table 2.2 demonstrates these trends by indicating that function (1) (simple logistic growth function) is statistically most significant. Function (3) (logistic growth function within a dynamic carrying capacity) follows function (1) with respect to its fittability. However, if we compare statistics of functions (1) and (3) in Table 2.2, we note that a dynamic carrying capacity is negligibly small (e.g. a_k : 4.141×10^-6) and carrying capacity of function (1) ($K = 5,226,506$) and function (3) ($K_K = 5,226,506$) has reached the same level. These support the former postulate that the diffusion process of refrigerators can be identified by the single logistic growth function.

(c) Similar to refrigerators, color TV sets have single function. However, their diffusion process is more complicated than refrigerators as they had substitution process with mono-color TV sets. In the initial diffusion process, color TV sets were co-evolution with monochrome TV sets. Since color broadcasting has become available in all TV broadcasting programs in 1973 in Japan, second diffusion emerged which is substantial diffusion process in a competitive market. Thus, diffusion process of color TV sets is typical bi-logistic growth. Table 2.2 demonstrates this trend by indicating that function (2) is statistically most significant.

Contrary to these diffusion processes in refrigerators and color TV sets, the diffusion process of cellular telephones is most complicated as it has multi-functions. Although cellular telephones have young history on their diffusion than refrigerators and color TV sets, they have the highest IT density. Such high IT density enables cellular telephones create new functions during the course of their interactions with customers leading them to be multi-functions goods. Diffusion process of this type of innovation could be well modeled by logistic growth within a dynamic carrying capacity. Table 2.2 demonstrates this postulate by AIC. Although AIC supports this postulate and other statistics also demonstrate better fittability than function (1), other statistics except AIC demonstrates slightly less significant than function (2). This could be interpreted that the diffusion process of cellular telephones has not yet matured and still is in transition from bi-logistic growth to logistic growth within a dynamic carrying capacity.

(d) Among three innovative goods examined, cellular telephones have definitely the highest IT density and multi-functions. Their diffusion process is identified to logistic growth within a dynamic carrying capacity that represents such diffusion process as correlation of the interaction between innovations and institutions displays systematic change in their process of the growth and maturity. This demonstrates our hypothetical view that IT's specific features are formed through dynamic interaction with an institutional system.

2.2.3 Features of IT

As examined in Sect. 2.2, cellular telephones that contain the highest IT density as a crystal of mobile communications technology, one of the most representative and most popular information technology, was verified that its diffusion process, or behavior, well matches the logistic growth function within a dynamic carrying capacity. Consequently, (2.7) implies and Fig. 2.2 demonstrates that IT's epidemic behavior closely interrelates with continuous increase in number of potential users. It means that during the course of diffusion, IT interacts with individuals, organizations, and society as a whole, changes its behavior depending on institutions of that it interacts with, and extends potential users with its newly acquired features. This characterizes unique diffusion process of IT in that it alters carrying capacity or creates a new carrying capacity in the process of its diffusion, thereby acquires new specific features.

Figure 2.8 compares diffusion process of manufacturing technology and IT. Each time IT interacts with institutions, it effects institutions and potential users within them to change as well as acquires new features depending on institutions. Thus,

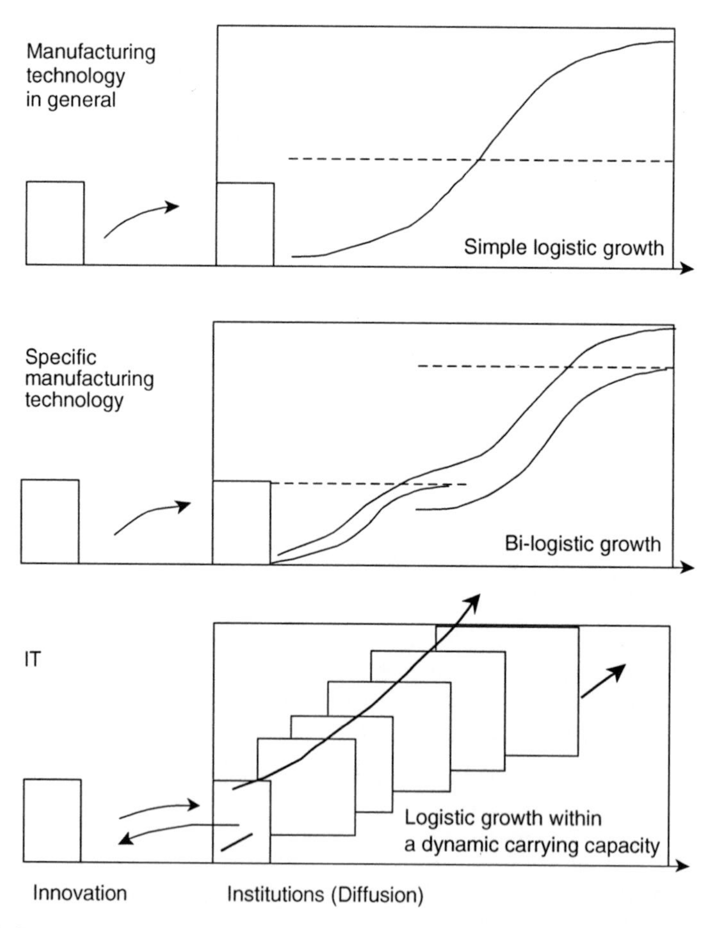

Fig. 2.8 General concept of technology diffusion process – comparison between manufacturing technologies and IT

IT's diffusion process is stimulated by interaction with institutions and institutional change is also stimulated by interaction with IT, leading to co-evolution of technology itself and institutions as well as constructing a virtuous cycle between rising technology value and increasing potential users.

By focusing on the above observed unique epidemic behavior of IT, the following specific features of IT with respect to interaction with institutions can be derived. Since IT behaves differently depending on the institutions of that it interacts with, whether a nation can fully exploit the benefits of IT greatly depends on the nation's institutional elasticity to the following features of IT:

- *Disseminative*

As the logistic growth function within a dynamic carrying capacity itself represents, dynamic evolution of a carrying capacity can be directly connected to a disseminative feature of IT. As the famous Moore's Law shows,[8] technological development of IT is very rapid. This rapid development of the technology together with network externalities enable IT related products and services to disseminate rapidly. The Economic Planning Agency [21] points out that appropriately judging the surrounding environment and quickly commercializing new products and services are crucial to survive in an information society. To make the best use of the disseminative nature of IT, organizations are required to make decisions quickly and react elastic to changing environments. Increasing significance of global technology spillover should be realized in a similar context [22].

- *Interactive*

The leading player of the IT industry is now shifting from personal computers to networks [23]. With the development of advanced and global networks, such as the Internet and mobile communications, more and more people can communicate and exchange information without being restricted by time and distance. Each time people use networks to communicate or exchange information, they actually interact with IT, and phenomena of network externalities that push up the carrying capacity rise the value of networks through the interaction between people and IT. Accordingly, interactiveness of IT is an important feature to explain IT's unique behavior.

Inside organizations, the interactive nature of IT improves efficiency of decision making process and induces structural transformation of organizations from hierarchical to network-type [4]. In order to exploit the potential benefits of IT, a reorganization of work that introduces new work practices is necessary [2]. In this sense, sticking to a conservatively hierarchical organization hinders efficient communication within an organization in spite of the interactive environment provided by IT.

- *Co-evolutional*

As Figs. 2.2 and 2.8 illustrate, features of IT and institutions evolve together during the course of their interaction that derives co-evolutional feature of IT. With its disseminative feature, IT diffuses as a social infrastructure and transforms economic,

[8] Chip capacity doubles every 18 months.

social, and cultural system of nations. In an information society, where IT functions as social infrastructure, growth depends more than ever on responding to the changing demands of the workplace and society more broadly [2]. In this context, with the indigenous nature of IT as a social infrastructure, the most effective way to maximize the benefits of IT should be the spontaneous evolution of the society itself as the technological development proceeds.

In addition to these three features, the following two features that are more or less common to all technologies are conspicuous in the behavior of IT.

– *Global*

IT, especially a network technology, realizes global information exchange independent of time, distance, and even borders. The global information exchange enables incessant flow of unknown but maybe effective cultures and services, as well as global procurement of goods and human resources which leads to reduction in production costs. To make the best use of this global nature, high adaptability to changing environment and ability to absorb heterogeneous cultures are required. The melting pot of the US well matches this heterogeneous environment and enjoys the benefits of globalization brought by IT [24].

– *Invisible*

IT should be referred to as cross-industrial technology [21] and would play its role as a social infrastructure that invisibly supports social and economic activities. The Digital Economy 2000 [6] claims that IT innovations can be applied across the economy and throughout the economic process: IT provides new ways of managing and using a resource that is common to every sector and aspect of economic life. On this invisible infrastructure, various services are expected to be created across all industries. Because of the unlimited potentiality of IT, creativity and entrepreneurship play a key role to create new businesses and achieve high growth using this infrastructure.

Finally, as network externalities contribute significantly to enhance IT's carrying capacity, increasing returns to scale phenomena are observed in high IT intensity industry. Table 2.3 demonstrates these phenomena by measuring SCE (scale of economies) [25] in major Japanese manufacturing industry over the period 1976–1998.

Table 2.3 Comparison of SCE in Japan's manufacturing industry

	SCE (%)
General machinery	32.34
Electrical machinery	29.60
Transportation equipment	28.27
Precision instruments	17.53
Chemicals	−18.87
Primary metals	−21.94
Pulp and paper	−22.53
Metal products	−30.29
Ceramics	−47.10

As summarized in Table 2.3, high IT intensity sectors as general machinery, electrical machinery, transportation equipment, and precision instruments display positive SCE demonstrating increasing returns to scale while other sectors with relatively low IT intensity demonstrate decreasing returns to scale (see Appendix 3 for mathematical development of SCE).

2.3 Implications

As analyzed so far, since unique features of IT such as *disseminative*, *interactive*, *co-evolutional*, *global*, and *invisible* are formed during the course of interaction with institutional systems, institutional elasticity plays a significant role in inducing and diffusing IT as well as fully exploiting potential benefits of IT. If a nation's indigenous institution can react elastic to the advancement of IT, diffusion process of IT is accelerated, then that nation should be able to exploit potential benefits of IT, resulting in enhancing its international competitiveness.

In general, Japanese managerial activities and systems reflect its institutional characteristics. The centuries of isolation from foreign influences ("*sakoku*") during the Edo period (1603–1867) has meant that the Japanese population is culturally and ethnically more homogeneous than in most other countries [26]. This fairly homogeneous population, together with highly dense population in Japan, has contributed to develop unique features of the Japanese organizational and behavioral norms such as group orientation and feeling comfortable to "be the same" as others (neighbors) are McMillan [27] concisely characterizes the Japanese as consensual, highly stable, homogeneous, disciplined, and long term oriented.

During the "catching-up" period up to the end of the 1980s when manufacturing technology was considered as a core technology of an industrial society, Japanese business management such as lifetime employment, seniority system, and *keiretsu* well matched the nation's institutions and successfully established the feeling of "family ties" that led the nation to achieve high economic growth. Japanese manufacturers intensely developed products with their own in-house technology since customers were most interested in the quality of products, not so much in compatibility among products of different manufacturers. In order to assure quality, firms preferred in-house procurement of manufacturing parts, or relied on their *keiretsu* companies that reflected Japanese long term orientation. In other words, Japanese firms used individual language that consequently excluded entities outside the family.

By contrast, IT enables global information exchange, thus induces global procurement of goods and mobility of human resources. Facing this new paradigm emerged with a dramatic advancement of IT in the 1990s, Japan can no longer depend on well-tried, low-risk paths and other benefits available to a country undergoing "catching-up" [26]. Furthermore, Japanese indigenous features such as homogeneousness and preferring high stability together with the existence of individual language peculiar to firms cannot react elastic to disseminative, interactive, co-evolutional, global, and invisible features of IT. Consequently, Japan's institutional system, which performed efficiently in the 1980s, is not efficient any more in an information society, and even hinders exploiting potential benefits of IT.

On the other hand, as MacRae [24] argued, the melting pot of the US makes the nation a great generator of new ideas, cultivates frontier spirit, and enhances flexibility to accept heterogeneous cultures, thus induces positive effects as a result of interaction with the above mentioned unique features of IT. In addition, the heterogeneous environment of the US stimulated the nation to establish standard language with that people or organizations can communicate implicitly. These features of the US institutional system did not perform effectively for an industrial society where steady and incremental advance was most appreciated.

Figure 2.9 summarizes how Japanese institutional systems and those of the US performed as the paradigm shift occurred from an industrial society to an information society. As analyzed above, though Japanese institutional systems were effective to the paradigm of an industrial society, they cannot be effectively applied to the new paradigm of an information society. Adversely, the US systems, which were ineffective in the paradigm of the 1980s, became pretty effective to the paradigm of the 1990s.

OECD [2] reported uneven trend growth of GDP per capita in OECD countries over the past decade compared with the 1980s. It described that trend growth in the 1990s was higher than in the 1980s in countries such as Australia, Canada, and the United States while growth declined markedly in such areas as Japan, Switzerland and Korea. Although there must be number of factors to explain these divergences,

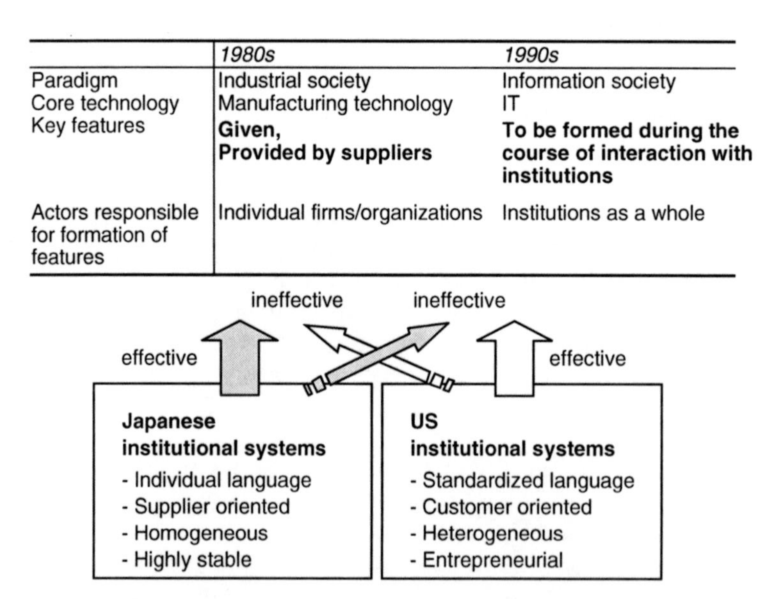

Fig. 2.9 Comparison of effectiveness between Japanese institutional systems and the US institutional systems under the paradigm shift

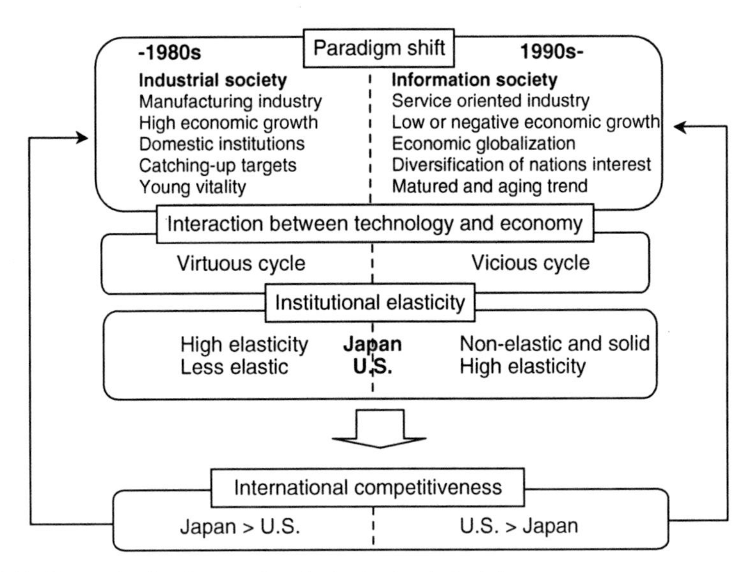

Fig. 2.10 Scheme leading Japan to lose its institutional elasticity

one of the factors should be attributed to Japan's solid institutional elasticity against unique features of IT that have been highlighted through dynamic interaction with institutional systems.

Figure 2.10 illustrates scheme leading Japan to losing its institutional elasticity by comparing the US system which indicates that, contrary to the dual virtuous cycle up to the end of the 1980s, Japan has been suffering from a dual vicious cycle.

As described above, during the period of an industrial society initiated by manufacturing industry, Japan's domestic institutions based on young vitality functioned efficiently towards "catching-up" target leading to high economic growth. In the 1990s, Japan's economy clearly contrasted with the preceding decades. Facing a new paradigm characterized by a shift to an information society initiated by service oriented industry, globalization, diversification of nations interest, aging trend, and subsequent low, zero or negative economic growth, Japan's traditional institutions did not function efficiently as they did in preceding decades.

Consequently, a virtuous cycle between institutional elasticity and economic development changed to a vicious cycle between non-elastic institutions and economic stagnation. This vicious cycle resulted in losing Japan's international competitiveness that reacted further economic stagnation. Thus, Japan has been facing a dual vicious cycle leading to a solid institutional elasticity.

2.4 Conclusion

In light of the understanding that effective utilization of potential benefits of dramatic advancement of IT in an information society will differ greatly depending on the nation, particularly on their institutional elasticity and that this can be attributed to the specific features of IT which performs its function in connection with institutional systems, this chapter attempts to derive specific IT features by focusing on the unique diffusion process, in other words epidemic behavior of IT.

An empirical analysis on the diffusion process of innovative goods in Japan was conducted taking refrigerators, color TV sets and cellular telephones which represent innovative goods centered by manufacturing technology and IT, respectively. On the basis of the comparative analysis of epidemic behavior between IT and other technologies using the simple logistic growth function, bi-logistic growth function and logistic growth function within a dynamic carrying capacity, it was demonstrated that the specific features of IT are formed through dynamic interaction with an institutional system. In addition, certain specific features of IT characterized during the course of the interaction process and conspicuous in its unique epidemic behavior were identified including disseminative, interactive, co-evolutional as well as extremely invisible and global than technology in general. Furthermore, a mechanism of IT's contribution to increasing returns to scale was identified.

These analyses provided us significant insight that Japan's industrial and management system based on non-stylized management system unique to individual firms/organizations well functioned for innovation and diffusion of manufacturing technologies which supported industrial society. Furthermore, it has become evident that such non-stylized management system does not function well for innovation and diffusion of IT features of which are formed through dynamic interaction with an institutional system to which stylized management system is indispensable.

All these findings remind us the significance of the role of institutional elasticity in making full utilization of potential benefit of IT and also of the urgency of remediation of Japan's lost institutional elasticity. Thus, systems functions supportive to complement for remediation of the institutional elasticity with distinct features of IT would be crucial.

Appendix 1. Mathematical Development of Logistic Growth Function Within a Dynamic Carrying Capacity

Simple logistic growth function is expressed as follows:

$$\frac{\mathrm{d}f(t)}{\mathrm{d}t} = bf(t)\left(1 - \frac{f(t)}{K}\right). \tag{2.8}$$

Given that innovation itself and the number of potential users change through the diffusion of innovation, logistic growth function within a dynamic carrying capacity is expressed by (2.9) where the number of potential users, carrying capacity (K) in the epidemic function is subject to a function of time t.

$$\frac{\mathrm{d}f(t)}{\mathrm{d}t} = bf(t)\left(1 - \frac{f(t)}{K(t)}\right). \tag{2.9}$$

Equation (2.10) is obtained from (2.9):

$$\frac{\mathrm{d}f(t)}{\mathrm{d}t} + (-b)\,f(t) = \left(-\frac{b}{K(t)}\right)\{f(t)\}^2. \tag{2.10}$$

Equation (2.10) corresponds to the Bernoulli's differential equation expressed by (2.11):

$$\frac{\mathrm{d}y}{\mathrm{d}x} + V(x)y = W(x)y^n. \tag{2.11}$$

Accordingly, (2.10) can be transformed to the linear differential equation expressed by (2.12).

$$\frac{\mathrm{d}z(t)}{\mathrm{d}x} + bz(t) = \frac{b}{K(t)} \text{ where } z(t) = \frac{1}{f(t)} \tag{2.12}$$

The solution for a linear differential (2.13) can be obtained as (2.14):

$$\frac{dy}{dt} + P(x)y = Q(x) \tag{2.13}$$

$$y = \exp\left(-\int P(x)dx\right)\cdot\left\{\int\left(Q(x)\cdot\exp\left(\int P(x)dx\right)\right)dx + c\right\} \tag{2.14}$$

Accordingly, the solution for (2.12) can be expressed as follows:

$$\begin{aligned} z(t) &= \exp(-\textstyle\int b\,\mathrm{d}t)\cdot\left\{\int\left(\frac{b}{K(t)}\exp(\textstyle\int b\,\mathrm{d}t)\right)\mathrm{d}t + c_1\right\} \\ &= \exp(-bt)\cdot\left\{b\int\left(\frac{1}{K(t)}\exp(bt)\right)dt + c_1\right\} \end{aligned} \tag{2.15}$$

$$\frac{1}{f(t)} = \exp(-bt)\cdot\left\{b\int\left(\frac{\exp(bt)}{K(t)}\right)\mathrm{d}t + c_1\right\}. \tag{2.16}$$

Assume that a carrying capacity $K(t)$ increases sigmoidally, $K(t)$ is expressed as follows:

$$K(t) = \frac{K_K}{1 + a_K\exp(-b_K t)}. \tag{2.17}$$

By substitution (2.17) for $K(t)$ in (2.16), (2.18) is obtained:

$$\frac{1}{f(t)} = \left\{b\int\left(\frac{\exp(bt)}{K_K/(1 + a_K\exp(-b_K t))}\right)\mathrm{d}t + c_1\right\}\exp(-bt), \tag{2.18}$$

where

$$
\begin{aligned}
&\int\left(\frac{\exp(bt)}{K_K/(1+a_K\exp(-b_Kt))}\right)\mathrm{d}t\\
&=\frac{1}{K_K}\int\{\exp(bt)+a_K\exp((b-b_K)t)\}\mathrm{d}t\\
&=\frac{1}{K_K}\left\{\int\exp(bt)\,\mathrm{d}t+\int a_K\exp((b-b_K)t)\,\mathrm{d}t\right\}\\
&=\frac{1}{K_K}\left\{\frac{1}{b}\exp(bt)+\frac{a_K}{b-b_K}\exp((b-b_K)t)\right\}+c_2. \qquad (2.19)
\end{aligned}
$$

Accordingly, $f(t)$ can be developed as follows:

$$
\begin{aligned}
\frac{1}{f(t)}&=b\left\{\frac{1}{K_K}\left\{\frac{1}{b}\exp(bt)+\frac{a_K}{b-b_K}\exp((b-b_K)t)\right\}+c_2+c_1\right\}\cdot\exp(-bt),\\
\frac{1}{f(t)}&=\frac{1}{K_K}\left\{1+\frac{b\cdot a_K}{b-b_K}\exp(-b_Kt)+c_3\exp(-bt)\right\},\\
\frac{1}{f(t)}&=\frac{1}{K_K}\left\{1+c_3\exp(-bt)+\frac{b\cdot a_K}{b-b_K}\exp(-b_Kt)\right\}, \qquad (2.20)\\
f(t)&=\frac{K_K}{1+a\exp(-bt)+\frac{b\cdot a_K}{b-b_K}\exp(-b_Kt)}. \qquad (2.21)
\end{aligned}
$$

Appendix 2. Data Construction and Sources

In order to analyze the diffusion process of innovative products in Japan, domestic shipment (shipment for domestic demand) was regarded the most desirable indicator since it reflects the demand of customers for those products. However, since domestic shipment data was only available for color TV sets, we used production volumes for refrigerators and cellular telephones by making due adjustment for export and import balance.

A.2.1 Refrigerators

Annual shipment volume of refrigerators from the year 1966 to 1999 were obtained from the "Report on Machinery Statistics" conducted by the Ministry of International Trade and Industry (MITI).[9] Since the ratio of imports and exports

[9] MITI renamed the Ministry of Economy, Trade and Industry on January 6, 2001 under the structural reform of the Japanese government.

of refrigerators to their shipment as a whole has not been changing greatly, annual shipment volume was used for the analysis.

A.2.2 Color TV Sets

Annual domestic shipment volume of color TV sets from the year 1966 to 2000 were obtained from the Survey of Japan Electronics and Information Technology Industries Association (JEITA, annual issues).

A.2.3 Cellular Telephones

Since the ratio of imports and exports of cellular telephones to their domestic production volume as a whole has been changing and somewhat significant, the volume of imports and exports were considered for data construction. Quarterly production volume of cellular telephones from the year 1993 to 2000 was obtained from the "Report on Machinery Statistics" conducted by MITI, and quarterly volume of imports and exports from the year 1996 to 2000 was obtained from the "Trade Statistics" conducted by the Ministry of Finance. Although the volume of imports and exports of cellular telephones over the period of 1993–1995 was not available, since the ration of imports and exports to production is stable and relatively small before 1996, it was estimated by multiplying the same ratio of the year 1996 to production volumes.

Appendix 3. Mathematical Development of SCE

In order to measure SCE, the following production function was used:

$$\begin{aligned} V &= F(L,K,I,T,t) \\ &= F\{(L,I_l),(K,I_k),T,t\} \\ &= Ae^{\lambda t}\left(L^{\alpha_1}\cdot I_l^{\alpha_2}\right)\left(K^{\beta_1}\cdot I_k^{\beta_2}\right)T^{\gamma}, \end{aligned} \tag{2.22}$$

where A, scale coefficient; L, labor; K, capital; I, IT production factor; I_l, IT labor; I_k, IT capital; T, technology stock; and t, time trend. Duplication among each production element was deducted.

IT production factor was constructed using the data from the Ministry of International Trade and Industry's "Current Status of Japanese Information Processing," which referred the "Survey on Information Processing in Japan" by Japan

Table 2.4 IT related investment

Labor cost		Outsourced personal expenses, education and training cost, personal expenses, service charge, etc.
Capital cost	Hardware	Depreciation cost, rent fee, lease fee, installation charge, maintenance charge
	Software	Use charge, purchase cost, programming charge, consignment cost, machine rent charge, calculation consignment cost, data input charge
	Network	Network charge, network subscription charge, online service charge

Information Processing Development Center. "Capital Matrix of the Input–Output Tables" was also used to supplement the IT related investment that is not covered by the Survey. The resultant IT production factor is explained by the IT related investment listed in Table 2.4.

In order to analyze SCE using the production function given by (2.22), incorporation ability of technology spillovers should be measured as follows in light of the active spillover characteristics of IT among industries [22]:

$$I = I_{\mathrm{i}} + Z_{\mathrm{IT}}\, I_{\mathrm{s}} \tag{2.23}$$

$$Z_{\mathrm{IT}} = \frac{1}{1+\frac{\Delta I_{\mathrm{s}}/I_{\mathrm{s}}}{\Delta I_{\mathrm{i}}/I_{\mathrm{i}}}} \cdot \frac{I_{\mathrm{i}}}{I_{\mathrm{s}}}, \tag{2.24}$$

where I_{i}, own IT stock; I_{s}, potential IT spillover; and Z_{IT}, IT assimilation capacity.

By introducing incorporated spillovers of labor and capital respectively, the production function given by the (2.22) can be described as follows:

$$V = A\, \mathrm{e}^{\lambda t} \left\{ L^{\alpha_1} \left(I_{li} + Z_{il} I_{ls}\right)^{\alpha_2} \right\} \left\{ K^{\beta_1} \left(I_{ki} + Z_{ik} I_{ks}\right)^{\beta_2} \right\} T^{\gamma}. \tag{2.24$'$}$$

Christensen and Greene [25] defined SCE as follows:

$$\mathrm{SCE} = 1 - \frac{\partial \ln C}{\partial \ln y}, \tag{2.25}$$

where C, total cost; and y, real output.

If the increase in total cost is less than 1% while the output increases by 1%, SCE is greater than 0, that is scale economy works. If SCE $= 0$, it means constant returns to scale, and SCE < 0 indicates diminishing returns.

By using the production function (2.22$'$), where technology-related factors are deducted from L and K to avoid duplication, elasticity of each factor of production is expressed as follows that correspond to the ratio of costs:

$$\begin{aligned}
\alpha_1 &= \frac{\text{GLC}}{\text{GDP}} = \frac{P_l \cdot L}{P_v \cdot V} \\
\alpha_2 &= \frac{\text{GILC}}{\text{GDP}} = \frac{P_{il} \cdot I_l}{P_v \cdot V} \\
\beta_1 &= \frac{\text{GCC}}{\text{GDP}} = \frac{P_k \cdot K}{P_v \cdot V} \\
\beta_2 &= \frac{\text{GICC}}{\text{GDP}} = \frac{P_{ik} \cdot I_k}{P_v \cdot V} \\
\gamma &= \frac{\text{GTC}}{\text{GDP}} = \frac{P_t \cdot T}{P_v \cdot V}
\end{aligned} \tag{2.26}$$

Considering that GDP = GLC + GCC + GTC + GIC and by substituting GILC + GICC with GIC, (2.27) is obtained:

$$\alpha_1 + \alpha_2 + \beta_1 + \beta_2 + \gamma = 1. \tag{2.27}$$

From (2.26), the following equation is obtained:

$$\begin{aligned}
\alpha_1 &= \frac{\text{GLC}}{\text{GDP}} = \frac{P_l \cdot L}{P_v \cdot V} & \text{GDP} &= \frac{P_l \cdot L}{\alpha_1} \\
\alpha_2 &= \frac{\text{GILC}}{\text{GDP}} = \frac{P_{il} \cdot I_l}{P_v \cdot V} & \text{GDP} &= \frac{P_{il} \cdot I_l}{\alpha_2} \\
\beta_1 &= \frac{\text{GCC}}{\text{GDP}} = \frac{P_k \cdot K}{P_v \cdot V} \quad \Rightarrow & \text{GDP} &= \frac{P_k \cdot K}{\beta_1} \\
\beta_2 &= \frac{\text{GICC}}{\text{GDP}} = \frac{P_{ik} \cdot I_k}{P_v \cdot V} & \text{GDP} &= \frac{P_{ik} \cdot I_k}{\beta_2} \\
\gamma &= \frac{\text{GTC}}{\text{GDP}} = \frac{P_t \cdot T}{P_v \cdot V} & \text{GDP} &= \frac{P_t \cdot T}{\gamma}
\end{aligned} \tag{2.26'}$$

Given constant output to the production function (2.22′), total cost (C) is obtained by minimizing cost under constant price of production factors:

$$\begin{aligned}
C = {} & V^{\frac{1}{\alpha_1+\alpha_2+\beta_1+\beta_2+\gamma}} (\alpha_1 + \alpha_2 + \beta_1 + \beta_2 + \gamma) \left(\frac{1}{A\, \mathrm{e}^{\lambda t}}\right)^{\frac{1}{\alpha_1+\alpha_2+\beta_1+\beta_2+\gamma}} \\
& \times \left(\frac{P_l}{\alpha_1}\right)^{\frac{\alpha_1}{\alpha_1+\alpha_2+\beta_1+\beta_2+\gamma}} \left(\frac{P_{il}}{\alpha_2}\right)^{\frac{\alpha_2}{\alpha_1+\alpha_2+\beta_1+\beta_2+\gamma}} \left(\frac{P_k}{\beta}\right)^{\frac{\beta}{\alpha_1+\alpha_2+\beta_1+\beta_2+\gamma}} \\
& \times \left(\frac{P_{ik}}{\beta_2}\right)^{\frac{\beta_2}{\alpha_1+\alpha_2+\beta_1+\beta_2+\gamma}} \left(\frac{P_t}{\gamma_2}\right)^{\frac{\gamma}{\alpha_1+\alpha_2+\beta_1+\beta_2+\gamma}}
\end{aligned} \tag{2.28}$$

(2.28) is obtained by substitution (2.26′) in (2.29).

$$
\begin{aligned}
\ln C &= \ln\{C(V, P_l, P_{il}, P_k, P_{ik}, P_t)\} \\
&= \frac{\alpha_1+\alpha_2+\beta_1+\beta_2+\gamma}{\alpha_1+\alpha_2+\beta_1+\beta_2+\gamma} \ln \text{GDP} \\
&= \frac{\alpha_1}{\alpha_1+\alpha_2+\beta_1+\beta_2+\gamma} \ln \text{GDP} + \frac{\alpha_2}{\alpha_1+\alpha_2+\beta_1+\beta_2+\gamma} \ln \text{GDP} \\
&+ \frac{\beta_1}{\alpha_1+\alpha_2+\beta_1+\beta_2+\gamma} \ln \text{GDP} + \frac{\beta_2}{\alpha_1+\alpha_2+\beta_1+\beta_2+\gamma} \ln \text{GDP} \\
&+ \frac{\gamma}{\alpha_1+\alpha_2+\beta_1+\beta_2+\gamma} \ln \text{GDP} + \ln(\alpha_1+\alpha_2+\beta_1+\beta_2+\gamma)
\end{aligned}
\tag{2.29}
$$

SCE, defined by (2.25) can be obtained from (2.28):

$$
\text{SCE} = 1 - \frac{1}{\alpha_1+\alpha_2+\beta_1+\beta_2+\gamma}. \tag{2.30}
$$

Accordingly, coefficients $\alpha_1, \alpha_2, \beta_1, \beta_2, \gamma$ are estimated from (2.26) as follows under the assumption of cost minimization:

$$
\begin{aligned}
\frac{\alpha_1}{\beta_2} &= \frac{\text{GLC}}{\text{GICC}} \\
\frac{\alpha_2}{\beta_2} &= \frac{\text{GILC}}{\text{GICC}} \\
\frac{\beta_1}{\beta_2} &= \frac{\text{GCC}}{\text{GICC}} \\
\frac{\gamma}{\beta_2} &= \frac{\text{GTC}}{\text{GICC}}
\end{aligned}
\tag{2.31}
$$

Average of coefficients ratios are obtained from (2.32) using (2.31).

$$
\begin{aligned}
\widehat{\left(\frac{\alpha_1}{\beta_2}\right)} &= \exp\left[\frac{1}{n}\sum \frac{\text{GLC}}{\text{GICC}}\right], \\
\widehat{\left(\frac{\alpha_2}{\beta_2}\right)} &= \exp\left[\frac{1}{n}\sum \frac{\text{GILC}}{\text{GICC}}\right], \\
\widehat{\left(\frac{\beta_1}{\beta_2}\right)} &= \exp\left[\frac{1}{n}\sum \frac{\text{GCC}}{\text{GICC}}\right], \\
\widehat{\left(\frac{\gamma}{\beta_2}\right)} &= \exp\left[\frac{1}{n}\sum \frac{\text{GTC}}{\text{GICC}}\right],
\end{aligned}
\tag{2.32}
$$

where n denotes number of samples.

Equation (2.34), a Cobb–Douglas production function can be obtained under the assumption of cost minimization by calculating a new variable ($\hat{Z}$) with the estimate of (2.32) as below:

$$\widehat{Z} \equiv \ln(I_k)+\left(\frac{\hat{\alpha_1}}{\beta_2}\right)\ln(L)+\left(\frac{\hat{\alpha_2}}{\beta_2}\right)\ln(I_l)+\left(\frac{\hat{\beta_1}}{\beta_2}\right)\ln(K)+\left(\frac{\hat{\gamma}}{\beta_2}\right)\ln(T), \quad (2.33)$$

$$\ln(V) = \ln A + \hat{\beta_2}\cdot\hat{Z}+\lambda t. \quad (2.34)$$

Since $\hat{\beta_2}$ is obtained by (2.34), we can calculate the estimators of $\alpha_1,\alpha_2,\beta_1,\gamma$ by multiplying $(\alpha_1\hat{/}\beta_2),(\alpha_2\hat{/}\beta_2),(\beta_1\hat{/}\beta_2),(\gamma\hat{/}\beta_2)$ by $\hat{\beta_2}$. Based on the results of $\alpha_1,\alpha_2,\beta_1,\beta_2,\gamma$, we can measure SCE by (2.31).

References

1. P.M. Romer, Beyond Classical and Keynesian Macroeconomic Policy, Policy Options (1994) 15–21
2. OECD, The New Economy: Beyond the Hype, Final Report on the OECD Growth Project (OECD, Paris, 2001)
3. OECD, Special Issue on Information Infrastructures, STI Review (OECD, Paris, 1997)
4. Telecommunications Council, Japan, The Info-communications Vision for the 21st Century (Telecommunications Council for the Minister of Posts and Telecommunications, Tokyo, 2000)
5. V.W. Ruttan, Technology, Growth, and Development – An Induced Innovation Perspective (Oxford University Press, New York, 2001)
6. US DOC, Digital Economy 2000 (DOC, Washington, DC, 2000)
7. F. Cairncross, The Death of Distance (Harvard Business School Press, Boston, 1997)
8. E.M. Rogers, Diffusion of Innovations (The Free Press of Glencoe, New York, 1962)
9. Z. Griliches, Hybrid corn: an explanation in the economics of technical change, Econometrica 25, No. 4 (1957) 501–522
10. P.S. Meyer, Bi-logistic growth, Technological Forecasting and Social Change 47, No. 1 (1994) 89–102
11. E. Mansfield, Intrafirm rates of diffusion of an innovation, The Review of Economics and Statistics 45, No. 4 (1963) 348–359
12. E. Mansfield, Industrial research and technological innovation: an econometric analysis (Longman, London, 1969)
13. J.S. Metcalfe, The Diffusion of Innovation in the Lancashire Textile Industry, Manchester School of Economics and Social Studies 2 (1970) 145–162
14. K. Norris and J. Vaizey, The Economics of Research and Technology (George Allen & Unwin, London, 1973)
15. C. Marchetti, On strategies and fate, in Hafele et al. (ed.), Second Status Report on the IIASA Project on Energy Systems 1975 (IIASA, Laxenburg, Austria, 1976) 203–217
16. R. Coombs, P. Saviotti and V. Walsh, Economics and Technological Change (Macmillan Publishers Ltd., London, 1987)
17. S.M. Oster, Modern Competitive Analysis (Oxford University Press, New York, 1994)
18. T.C. Schelling, Social Mechanisms and Social Dynamics, in P. Hedstrom and R. Swedberg (eds.), Social Mechanisms: An Analytical Approach to Social Theory (Cambridge University Press, Cambridge, 1998) 32–44
19. P.S. Meyer and J.H. Ausbel, Carrying capacity: a model with logistically varying limits, Technological Forecasting and Social Change 61, No. 3 (1999) 209–214

20. Ministry of Posts and Telecommunications (MPT), Japan, White Paper 2000 on Communications in Japan (MPT, Tokyo, 2000)
21. Economic Planning Agency, White Paper on the Japanese Economy (Economy Planning Agency, Tokyo, 2000)
22. C. Watanabe, B. Zhu, C. Griffy-Brown and B. Asgari, Global Technology Spillover and Its Impact on Industry's R&S Strategies, Technovation 21, No. 5 (2001) 281–291
23. D.C. Moschella, Waves of power (AMACOM, New York, 1997)
24. H. MacRae, The world in 2020: power, culture and prosperity (Harvard Business School Press, Boston, 1995)
25. L.R. Christensen and W.H. Greene, Economies of scale in U.S. electric power generation, Journal of Political Economy 84, No. 4 (1976) 655–676
26. R. Aggarwal, The shape of post-bubble Japanese business: preparing for growth in the new millennium, in International Executive, vol. 38(1) (Wiley, New York, 1996) 9–32
27. C. McMillan, The Japanese Industrial System (Walter de Gruyter & Co., Paris, 1996)

Chapter 3
Institutional Elasticity as a Significant Driver of IT Functionality Development

Abstract Institutions drive innovation and stimulate broad diffusion. Not surprisingly, national systems of innovation are influenced by their institutional elasticity in response to changing market conditions. As nations move from industrial to information-based societies, a key factor governing institutional elasticity is how institutions integrate IT. Since IT functionality is intimately connected with institutional dynamics, unlike simple manufactured products such as refrigerators, IT's specific functionality is formed through dynamic interaction with institutional systems. Consequently, institutional elasticity is a critical factor in the functionality of IT and its subsequent self-propagating behavior.

This chapter analyzes one possible mechanism of IT functionality formation, with special attention to the interaction of the technology with institutional systems.

Reprinted from *Technological Forecasting and Social Change* 71, No. 7, C. Watanabe, R. Kondo, N. Ouchi, H. Wei and C. Griffy-Brown, Institutional Elasticity as a Significant Driver of IT Functionality Development, pages: 723–750, copyright (2004), with permission from Elsevier.

3.1 Introduction

Innovation, such as the development of new technologies, is undoubtedly recognized as a significant driving force in sustaining economic growth. However, as the OECD [1] shows, new technology itself represents only potential, and in order to exploit this potential, institutional change is necessary.

A rapid surge in IT around the world is inevitably forcing traditional societies to transform their socioeconomic structures. As the Telecommunications Council [2] noted, IT is hastening Japan's paradigm shift from an industrial society to an information society. However, if IT is merely introduced to replace part of the workforce so as to improve productivity, as was the case with automation, the full benefits of IT will not be utilized. This is because IT not only enhances task efficiency but also permeates an organization, or a society, to have an impact on structure and

C. Watanabe, *Managing Innovation in Japan: The Role Institutions Play in Helping or Hindering how Companies Develop Technology*, DOI: 10.1007/978-3-540-89272-4_3,

behavior. More precisely, IT waves, most recently exemplified by the growing popularity of the internet and mobile communications, are characterized by so-called network externalities[1] [3, 4] that *construct a self-propagating "virtuous cycle"* between the expanding number of users and the rising value of networks, rapidly diffusing throughout the social infrastructure to support socio-economic activities.

The OECD [1] analyzed the potential of IT to "automate" and "informate." It observed that more relative emphasis has been given to the "automate" option and that IT has often been introduced into organizations that were shaped independently of it. Thus, if an organization can reengineer itself to shift the balance away from the "automate" option towards the "communicate" option, it can become a learning institution with new sets of skills.

Accordingly, IT differs from other technologies since individuals, organizations, and societies use it to interact, and its features are formed dynamically through this interaction. Institutions behave differently depending on institutional elasticity. In other words, *the unique features of IT are formed during the course of interaction with institutional systems* [5, 6]. Following their induced innovation concept [7], Ruttan [3], in their postulate of "institutional innovation," suggest that "institutions are the social rules that facilitate coordination among people by helping them form expectations for dealing with each other" and also that "they reflect the conventions that have evolved in different societies regarding the behavior of individuals and groups." Consequently, *IT's unique features can be identified in the diffusion process with respect to organizations and societies.*

Research on *the diffusion of innovation* has been undertaken in broad fields and Rogers [8] attempted to systematize these works in his pioneer work in "Diffusion of Innovations." He defined "diffusion" as *the process by which an innovation is communicated through certain channels over time among the members of a social system.* He also identified four main elements in the diffusion of innovations: *innovation features, communication channels, time, and social system.* All of Rogers's postulates support our hypothetical view with respect to IT features formation process that IT's unique features can be identified in its diffusion process.

This diffusion process is actually quite similar to the contagion process of an epidemic disease. Grilliches [9] exhibits S-shaped growth. This process is well modeled by the *simple logistic growth function*, an epidemic function which was first introduced by Verhulst in 1845 [10]. Since the logistic growth function has proved useful in modeling a wide range of innovation processes, a number of studies applied this function in analyzing the diffusion process of innovations as well [9, 11–14].

While the simple logistic growth function treats the carrying capacity of a human system[2] as fixed, this capacity is actually subject to change [15, 16]. Among varieties of innovations, certain innovations alter their carrying capacity in the process of their diffusion which stimulates an increase in the number of potential users [17]. This increase, in turn, incorporates new features in the innovations. Meyer [10] extended the analysis of logistic functions to cases where dual processes operate by

[1] The value to a consumer of a product increases as the number of compatible users increases [22].

[2] Upper limit of the level of diffusion, see Sect. 2.2 for mathematical implications of this capacity.

referring to an example when cars first replaced the population of horses but then took on a further growth trajectory of their own. He postulated *bi-logistic growth* in an attempt to deal with the fact that this diffusion process that contains complex growth processes not well modeled by the single logistic.

In addition to the above diffusion processes exhibited by a single logistic growth and bi-logistic growth, in particular innovations, a correlation of the interaction between innovations and institutions displays systematic change in their process of the growth and maturity. This is typically the case of the diffusion process of IT in which *network externalities* [18] function to alter the correlation of the interaction which creates new features of the innovation, IT. In this case, the rate of adoption increases, usually exponentially until physical or other limits slow the adoption. Adoption is a kind of "social epidemic." Schelling [19] portrays an array of logistically developing and diffusing social mechanisms stimulated by these efforts. Meyer and Ausbel [20] introduced an extension of the widely-used logistic model of growth by allowing it for a sigmoidally increasing carrying capacity. They stressed that "evidently, new technologies affect how resources are consumed, and thus if carrying capacity depends on the availability of that resource, the value of the carrying capacity would change." This explains, the unique diffusion process of IT, which diffuses by altering the carrying capacity or creating a new carrying capacity in the process. Meyer and Ausbel proposed *logistic growth within a dynamic carrying capacity* to model this diffusion behavior.

Building on this foundation, this paper analyzes one possible mechanism of IT functionality formation in Japan, with special attention given to the interaction of the technology with institutional systems. Furthermore, this analysis helps to explain differences in institutional elasticity in Japan and US as well as the implications for policy reform.

Section 3.2 identifies specific features of IT by conducting comparative analysis of epidemic behavior between IT and other technologies. Section 3.3 extracts implications with respect to effectiveness of institutional systems for IT features formation. Section 3.4 briefly summarizes the key findings of the analysis and presents conclusions by discussing the significant role of institutional elasticity for the effective utilization of IT in techno-economic growth.

3.2 Identification of IT Features with Respect to Institutions

3.2.1 The Formation Process of Specific Features of Technology

As emphasized in numerous studies, IT is functioning as a driving force to transform the existing socioeconomic structure by permeating people's daily life, organizational activities, and society as a whole, hastening the paradigm shift from an industrial society to an information society [2,4,5,21].

Table 3.1 Comparison of features between manufacturing technology and IT

	1980s	1990s
Paradigm	Industrial society	Information society
Core technology	Manufacturing technology	IT
Key features	*Given, provided by suppliers*	*To be formed during the course of interaction with institutions*
Actors responsible for formation of features	Individual firms/organizations	Institutions as a whole
System structure	Optimization	Standardization

Table 3.1 compares features of the core technologies in the 1980s and in the 1990s. During the 1980s, developing excellent manufacturing technology was critical for firms to be successful in an industrial society. Manufacturing technology was developed by the supply side to provide end-users with products and was introduced to factories to replace part of the workforce for improving productivity. Like other technologies, features of manufacturing technology are established or programmed at the beginning and once it leaves the supply side, it does not change its basic use substantially during its dissemination. In this case, individual firms are responsible for forming features of technology.

With information technology (IT) development, especially increased electronic connectivity in the 1990s, socio-economic activities have been relying more on IT infrastructure. This has made business transactions and information exchange easier and cheaper leading to an expanding e-commerce market [4].

Unlike manufacturing technology, suppliers of IT are more concerned about compatibility. This is because IT products are often utilized as a communication tool and exist in a complex technological "web". On the other hand, any home appliances such as refrigerators or TV sets can be purchased without being constrained and effected by compatibility issues related to technologies people possess besides electricity. In this context, IT products are subject to network externalities. With computers and telephones, for example, the more people use compatible systems or the more people are on a network, the more valuable the system or the network becomes, thus attracting more potential users [3].

In short, IT strongly possesses a self-propagating feature that closely interacts with individuals, organizations, and society during the course of its diffusion and behaves differently depending on the institutions involved. These observations suggest that functionality is formed dynamically during the course of interaction with institutions. Furthermore, whether the potential benefits of IT can be exploited greatly depends on the nature of these institutions.

This formation process of IT features is actually quite similar to the contagion process of an epidemic disease. Based on this similarity and the hypothetical view expressed above, the following analysis explains functionality development in the context of this self-propagating behavior.

3.2.2 Analysis of Epidemic Behavior

3.2.2.1 Taxonomy of Epidemic Function

In line with the approaches introduced in Sect. 3.1 [10, 20], the following three functions were used for a comparative analysis of epidemic behaviors between IT and other technologies:

(1) *Simple logistic growth function*:

$$f(t) = \frac{K}{1 + a\exp(-bt)}, \tag{3.1}$$

where $f(t)$, number of adopters; a and b, coefficients; K, carrying capacity (ceiling of the adoptions of innovative goods); and t, time trend.

(2) *Bi-logistic growth function*:

$$f(t) = f_1(t) + f_2(t) = \frac{K_1}{1 + a_1 \exp(-b_1 t)} + \frac{K_2}{1 + a_2 \exp(-b_2 t)}, \tag{3.2}$$

where a_1, a_2, b_1 and b_2, coefficients; K_1 and K_2, carrying capacities; and t, time trend.

(3) *Logistic growth function within a dynamic carrying capacity*:

The epidemic function expressed by (3.1) assumes that the level of carrying capacity (K) is constant through the dissemination process of innovation. However, as reviewed in Sect. 3.1, in particular innovations, correlation of the interaction between innovation and institutions display a systematic change in their process of growth and maturity leading to the creation of a new carrying capacity in the process of its diffusion. In these innovations, the level of carrying capacity will be enhanced as their diffusion proceeds, and carrying capacity K in (3.1) should be treated as the following function:

$$\frac{\mathrm{d}f(t)}{\mathrm{d}t} = bf(t)\left(1 - \frac{f(t)}{K(t)}\right), \tag{3.3}$$

where $K(t)$ is also an epidemic function enumerated by (3.4).

$$K(t) = \frac{K_K}{1 + a_K \exp(-b_K t)}, \tag{3.4}$$

where K_K indicates carrying capacity (the ultimate upper limit).

The solution of a differential equation (3.3) under the condition (4) can be obtained as an equation (3.5).

$$f(t) = \frac{K_K}{1 + a\exp(-bt) + \frac{b \cdot a_K}{b - b_K}\exp(-b_K t)}, \tag{3.5}$$

where a, b, a_K and b_K, coefficients; K_K, carrying capacity; and t, time trend.

In case when $a_K = 0$, (3.5) is equivalent to (3.1). Thus, (3.5) is a general function of the epidemic behavior encompassing a simple logistic growth function.

The dynamic carrying capacity $K(t)$ can be expressed by (3.6) by transforming (3.3).

$$K(t) = f(t)\left(\frac{1}{1-\left(\mathrm{d}f(t)/\mathrm{d}t\right)/bf(t)}\right). \quad (3.6)$$

Equation (3.6) demonstrates that $K(t)$ increases together with the increase of $f(t)$ as time goes by. This implies that (3.5) exhibits logistic growth within a dynamic carrying capacity as it displays a systematic change.

From (3.6) the allowance between the diffusion level and its ceiling $(K(t)/f(t))$ can be enumerated by the following equation:

$$K(t)/f(t) = \left[1-\frac{1}{b}\left\{\left(\mathrm{d}f(t)/\mathrm{d}t\right)/f(t)\right\}\right]^{-1}. \quad (3.7)$$

Equation (3.7) suggests that the allowance increases as the diffusion rate $((\mathrm{d}f(t)/\mathrm{d}t)/f(t))$ increases and the value of coefficient b decreases.

3.2.2.2 Comparative Analysis of Epidemic Behavior

In order to verify the difference in the diffusion processes, particularly in trends in carrying capacity, between IT and other technologies, the diffusion patterns of (1) refrigerators, (2) fixed telephones, (3) Japanese word processors, (4) color TV sets, (5) personal computers, and (6) cellular telephones were analyzed by applying a logistic growth function within a dynamic carrying capacity.

These six innovative goods were chosen based on the dimensions illustrated in Fig. 3.1. As described previously, since IT's diffusion process is characterized by its self-propagating behavior with creating new functionality through interactions with institutional systems, two dimensions were introduced in order to distinguish IT intensive products from other manufacturing products: the degree of multi-functionality and the user's manageability of functionality.

Products positioned in the bottom-right part in Fig. 3.1 such as personal computers and cellular telephones are regarded as IT intensive products, while products with mono functionality and limited users manageability such as refrigerators and fixed telephones are regarded as representative products of manufacturing technology. Fixed telephones are somewhat different from cellular telephones in that they mainly provide a voice exchange function while cellular telephones enable both voice exchange and data transmission. Color TV sets which started with the mono-function televising or pushing information to the viewer has evolved since the introduction of BS digital broadcasting service in 2000. Color TVs gained additional functions such as permitting viewers broad options in accessing information and also participating in the contents of the programs. Thus, the technology in a limited way is shifting from mono-functionality to with multi-functionality. Japanese

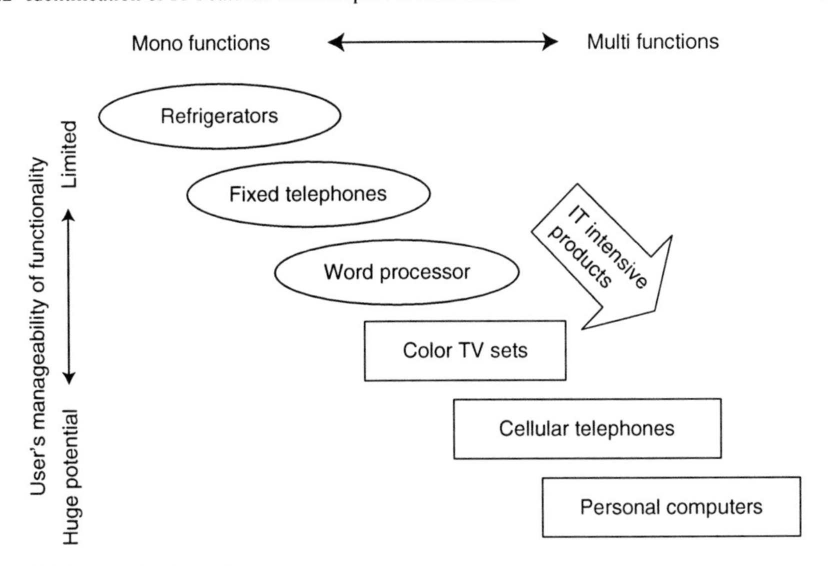

Fig. 3.1 Categorization of innovative goods

word processors once proliferated as a substitute for typewriters and can now be considered as transitional products between typewriters and personal computers.

The cumulative number of respective innovative goods was used in analyzing the diffusion patterns of these goods in Japan.[3]

Results are illustrated in Figs. 3.2–3.7.

Looking at Figs. 3.2–3.7 we note the following findings with respect to trends in carrying capacity of the respective innovative goods:

(1) *Refrigerators*:
The cumulative number of refrigerators has shown a rapid increase in the beginning of the 1960s supported by the economic boom called the "Iwato Boom", and continued to increase over the 1970s. While slightly stagnating, the cumulative number still maintained its steady increase. This is probably due to the innovation of freezers, a dramatic improvement in energy efficiency, and the innovation of refrigerators with large capacity. Since these innovations were in the scope of the same function, while refrigerators maintained a carrying capacity, the level has been nearly constant except for an increase in the 1950s and the early 1960s.

(2) *Fixed telephones*:
Similar to refrigerators, the cumulative number of fixed telephones has shown a rapid increase from the middle of the 1960s (lasting two decades) up to the middle of the 1980s. While slightly stagnating, the cumulative number still maintained its steady increase. This is due to the privatization of NTT and the succeeding liberalization of the sales of terminal receivers that brought about nicely designed telephones with convenient functions such as message recording and facsimile transmission or receiving, and effectively created new

[3] See Appendix for data construction and sources.

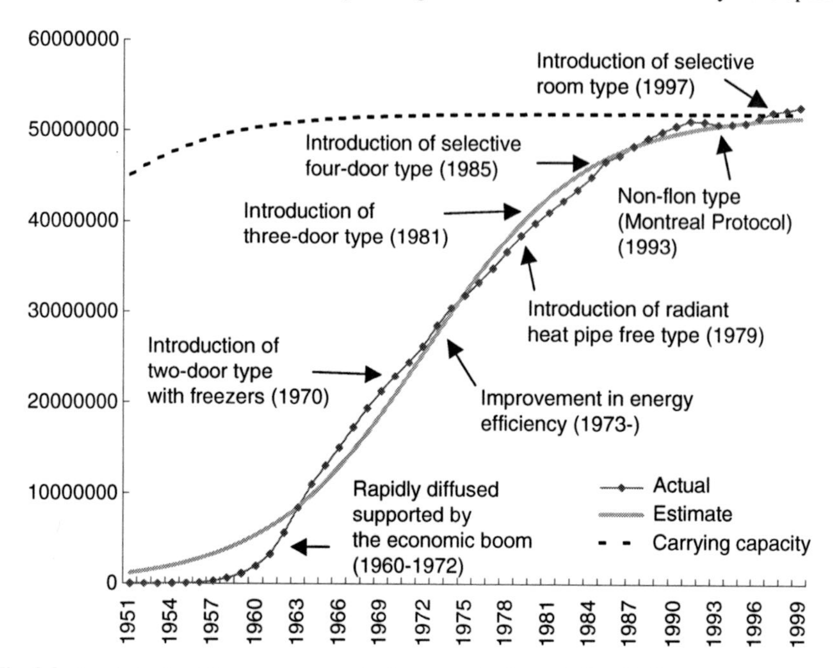

Fig. 3.2 Trends in the diffusion process of refrigerators in Japan (1951–1999)

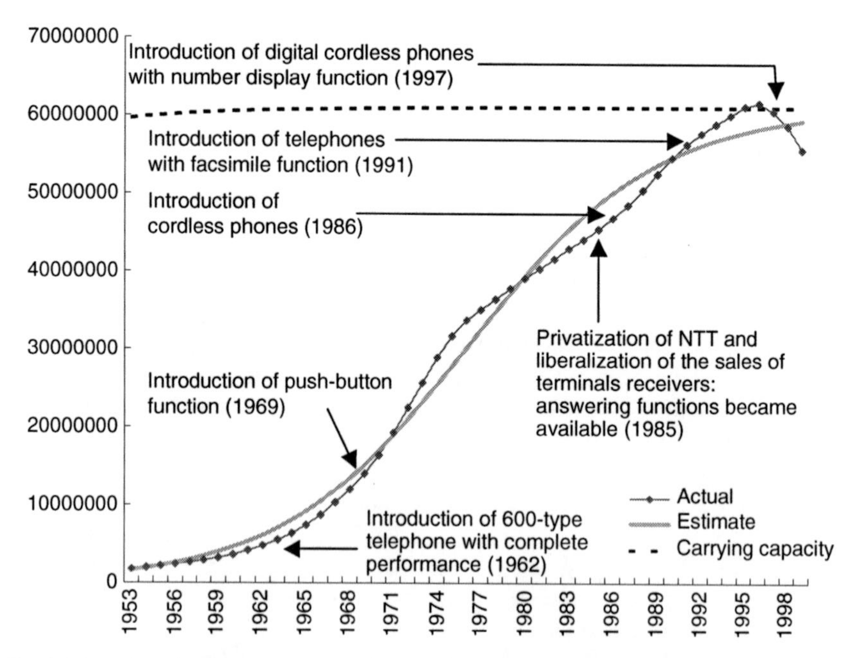

Fig. 3.3 Trends in the diffusion process of the fixed telephones in Japan (1953–1999)

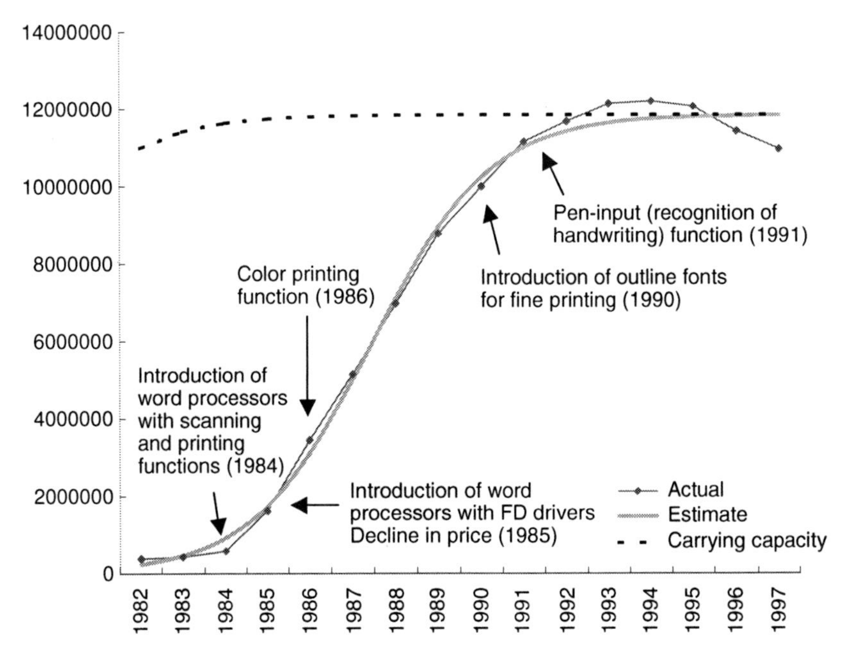

Fig. 3.4 Trends in the diffusion process of the Japanese word processors in Japan (1982–1997)

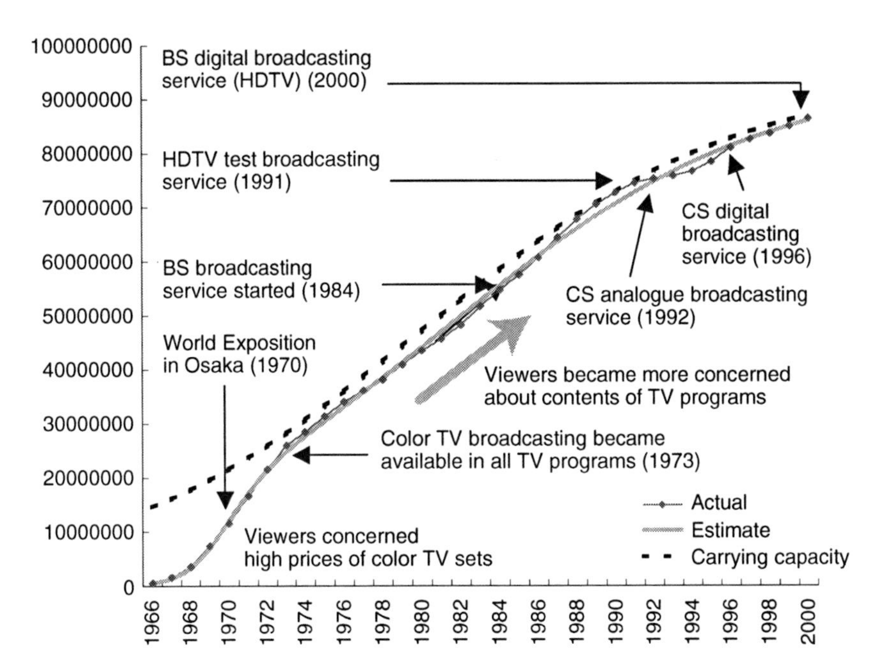

Fig. 3.5 Trends in the diffusion process of the color TV Sets in Japan (1966–2000)

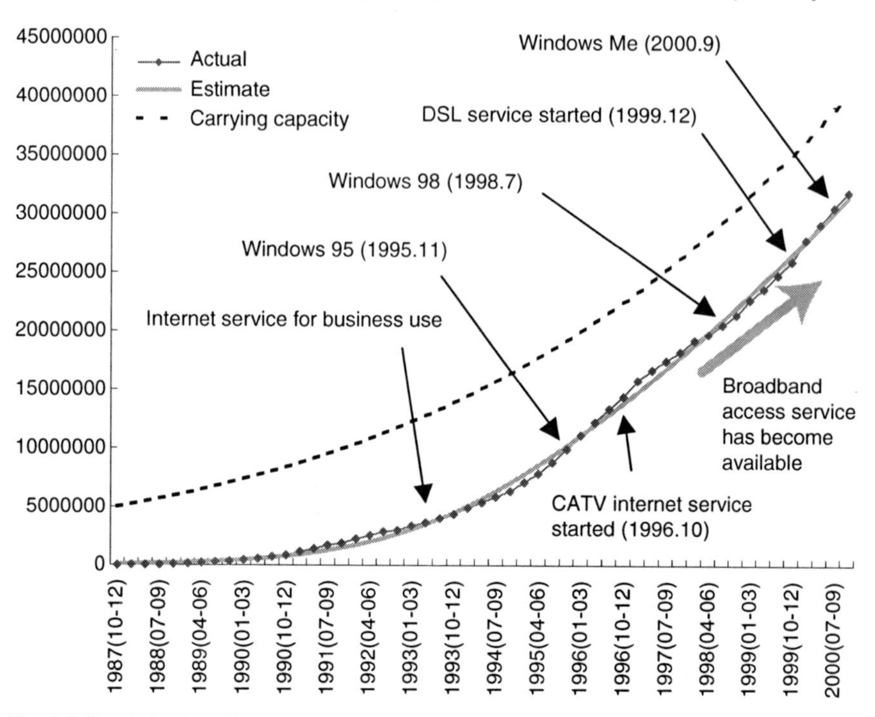

Fig. 3.6 Trends in the diffusion process of the personal computers in Japan (1987–2000)

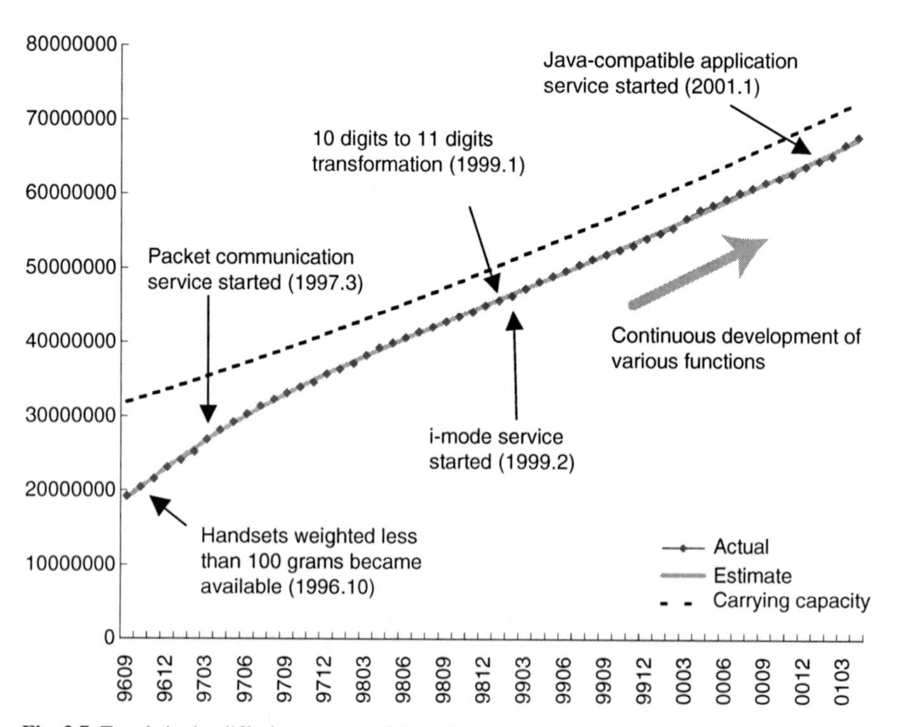

Fig. 3.7 Trends in the diffusion process of the cellular telephones in Japan (1996–2001)

demand. Similar to the case of refrigerators, since these innovations were in the scope of the same function, while fixed telephones maintained their carrying capacity, the level has been nearly constant except for a slight increase in the 1950s and to the middle of the 1960s.

(3) *The Japanese word processors*:
The cumulative number of Japanese word processors has shown a rapid increase in the limited period of the last half of the 1980s when the price reached a reasonable level for personal users. While dramatically declining from the beginning of the 1990s, the cumulative number still maintained a consistent increase. This is due to innovations such as a memory function, graphic and color processing functions, downsizing, and the fact that the technology maintained its niche function as a transitional product before being overtaken and substituted by personal computers. Although improved, the overall functionality has not made any substantial qualitative change. Thus, the carrying capacity level has been nearly constant except for a slight increase in the first half of the 1980s.

(4) *Color TV sets*
The cumulative number of color TV sets rapidly increased before color TV broadcasting became widely available in 1973. While this trend subsequently declined, it also maintained a consistent increase overall. This is due to successive innovations including the introduction of the BS broadcasting service in 1984 that afterwards started high definition test TV broadcasting service in 1991 and the CS (communications satellite) broadcasting service in 1992. These newly started services have provided viewers with new functionality including broad options in enjoying a variety of high-quality entertainment programs and clear HDTV images, accessing required information on demand and also interactively participating in TV programs. Therefore, the carrying capacity of color TV sets has increased as their cumulative number increased. However, the allowance between them has been limited over the period except the period before color TV broadcasting became available in all TV programs in 1973. This is due to the limitations of the new functions which were not necessarily evolutional ones.

(5) *Personal computers*
The cumulative number of personal computers has shown a constant increase with a higher increase from 1994 when advanced functions both hardware and software were broadly introduced into personal computers and the internet became widely available that enhanced the network externality of personal computers. Due to the self-propagating nature with respect to substantially new functions, their carrying capacity demonstrated a parallel path with an increase in the cumulative number. The broad "allowance" or distance between these trend lines occurred over the period 1991–1994 corresponding to the initial stagnation in the cumulative number of Japanese word processors. This may suggest that personal computers have been developing in a co-evolutionary way with the same producers switching production from word processors to personal computers. The recent development of broadband network access services such as cable TV internet and DSL seem to further stimulate the diffusion of personal computers.

(6) *Cellular telephones*[4]

Though cellular telephones have a relatively young history compared with that of the other innovative goods, the continuous development of smaller and lighter handsets with a variety of functions has made their diffusion process rather complicated and swift. One of the breakthroughs was NTT DoCoMo's introduction of i-mode service in February 1999 that enabled users to access the internet from their handsets. Since then, this kind of mobile internet access service has been dramatically expanding and the number of subscribers reached about 31.4 million as of February 2001.[5] Java-compatible handsets have now been on sale since January 2001, which is expected to induce a further increase in carrying capacity. Reflecting this structure, the cumulative number of cellular telephones has maintained a constant increase and their carrying capacity has increased in parallel with the increase in the cumulative number keeping a certain-fixed distance between the two trends. In contrast to the distance between carrying capacity and cumulative number in personal computers, this "allowance" is fixed demonstrating a striking self-propagating feature with respect to functionality development.

3.2.2.3 Interpretations

Table 3.2 compares the fit of the logistic growth function within a dynamic carrying capacity for the diffusion process of six innovative goods: refrigerators, fixed telephones, Japanese word processors, color TV sets, personal computers and cellular telephones.

Looking at Table 3.2, we note the following findings with respect to the behavior of a dynamic carrying capacity of these innovative goods:

(a) Table 3.2 demonstrates all indicators are statistically significant for six innovative goods examined except refrigerators' t-values on a and a_K.
(b) The adjusted R^2 demonstrates that the logistic growth function within a dynamic carrying capacity represents the actual diffusion behavior of six innovative goods in the market place.
(c) Parameters a_K for refrigerators, fixed telephones and Japanese word processors are extremely small values in comparison to the values for color TV sets, personal computers and cellular telephones which demonstrates that epidemic behaviors of the first three innovative goods are similar to the behavior of simple logistic growth while epidemic behaviors of the latter three innovative goods are with dynamic carrying capacity. In particular, the a_K for refrigerators is statistically insignificant and the value of this coefficient for fixed telephones is extremely small. These further demonstrate that the epidemic behaviors of refrigerators and fixed telephones are similar to typical simple logistic growth.

[4] Cellular telephones include PHS (personal handy-phone systems) and automobile phones as well as cell phones.

[5] http://www.tca.or.jp/

Table 3.2 Comparison of the fit of logistic growth function within a dynamic carrying capacity for the diffusion process of six innovative goods

	K_K	a	b	a_K	b_K	$adj.R^2$	DW
Refrigerators	51,884,200	31.793	0.177	0.181	0.175	0.999	0.09
	(121.66)	(0.96)	(29.46)	(0.17)	(26.17)		
Fixed telephones	60,948,330	4.177	0.155	0.026	0.155	0.997	0.18
	(56.85)	(12.60)	(23.14)	(10.41)	(23.11)		
Japanese word processors	11,849,440	1.207	0.722	0.163	0.721	0.997	0.75
	(78.27)	(1.10)	(17.40)	(3.51)	(17.24)		
Color TV sets	94,780,600	470.330	1.011	6.203	0.121	0.999	0.38
	(71.91)	(2.01)	(9.68)	(22.36)	(31.74)		
Personal computers	172,329,500	1,947.517	0.180	34.996	0.045	1.000	0.20
	(5.88)	(7.27)	(49.33)	(7.06)	(17.93)		
Cellular telephones	157,768,400	3.371	0.182	4.038	0.022	1.000	0.80
	(8.70)	(7.49)	(19.71)	(8.13)	(15.50)		

[a]Parameters are indicated in the following function:

$$f(t) = \frac{K_K}{1 + a\exp(-bt) + \frac{b \cdot a_K}{b - b_K}\exp(-b_K t)}.$$

[b]Figures in parentheses indicate t-value.

(d) Among the three innovative goods within a dynamic carrying capacity, the value of coefficient b for color TV sets is more than five times higher than personal computers and cellular telephones. This implies that while the diffusion of color TV sets has developed with dynamic carrying capacity, its level is approaching a ceiling or upper limit.
(e) In contrast to the diffusion of color TV sets, the allowances between diffusion level and the ceiling for personal computers and cellular telephones are still unlimited as the values of coefficients b are small enough and also their diffusion rates are high enough.
(f) All statistical interpretations correspond to observations illustrated in Figs. 3.2–3.7.

In order to further demonstrate the significance of a dynamic carrying capacity of these three innovative goods, Table 3.3 compares the fit of the three epidemic functions for the diffusion process of these innovative goods. Table 3.3 also compares the Akaike Information Criteria (AIC) of the three functions for the respective innovative goods.

Looking at Table 3.3, we note the following findings with respect to the identification of epidemic behavior for color TV sets, personal computers and cellular telephones:

(a) AIC demonstrates that the logistic growth function within a dynamic carrying capacity fits better than a simple logistic growth function for these three innovative goods.
(b) While AIC suggests that the logistic growth function within a dynamic carrying capacity demonstrates has a better fit than bi-logistic growth for cellular

Table 3.3 Comparison of the fit of three epidemic functions for the diffusion process of six innovative goods

Color TV sets								
(1)	K	a	b				$adj.R^2$	AIC
	86976,510	11.398	0.161				0.993	1.233×10^{13}
	(41.87)	(10.02)	(17.86)					
(2)	K_1	a_1	B_1	K_2	a_2	b_2	$adj.R^2$	AIC
	38,029,280	25.458	0.463	45,781,570	440.420	0.288	0.999	1.165×10^{12}
	(72.59)	(4.76)	(13.73)	(143.25)	(3.50)	(18.84)		
(3)	K_K	a	b	a_K	b_K		$adj.R^2$	AIC
	94,780,600	470.330	1.011	6.203	0.121		0.999	1.314×10^{12}
	(71.91)	(2.01)	(9.68)	(22.36)	(31.74)			
Personal computers								
(1)	K	a	b				$adj.R^2$	AIC
	39,716,890	137.844	0.115				0.999	3.517×10^{11}
	(29.10)	(12.08)	(33.32)					
(2)	K_1	a_1	b_1	K_2	a_2	b_2	$adj.R^2$	AIC
	26,403,230	183.897	0.144	26,403,250	2,668,028	0.263	0.999	7.564×10^{10}
	(31.18)	(12.55)	(37.33)	(31.18)	(2.80)	(37.56)		
(3)	K_K	a	b	a_K	b_K		$adj.R^2$	AIC
	172,329,500	1,947.517	0.180	34.996	0.045		0.999	1.250×10^{11}
	(5.88)	(7.27)	(49.33)	(7.06)	(17.93)			
Cellular telephones								
(1)	K	a	b				$adj.R^2$	AIC
	80824020	2.743	0.044				0.999	1.241×10^{12}
	(36.69)	(37.18)	(24.71)					
(2)	K_1	a_1	$\mathbf{b_1}$	K_2	a_2	b_2	$adj.R^2$	AIC
	39,420,150	1.278	0.114	39,420,160	41.825	0.082	0.999	6.140×10^{10}
	(78.66)	(53.53)	(31.65)	(79.41)	(19.94)	(55.48)		
(3)	K_K	a	b	a_K	b_K		$adj.R^2$	AIC
	157,768,400	3.371	0.182	4.038	0.022		0.999	6.007×10^{10}
	(8.70)	(7.49)	(19.71)	(8.13)	(15.50)			

[a]Parameters are indicated in the following functions, respectively:

(1) $f(t) = \frac{K}{1+a\exp(-bt)}$

(2) $f(t) = f_1(t) + f_2(t) = \frac{K_1}{1+a_1\exp(-b_1 t)} + \frac{K_2}{1+a_2\exp(-b_2 t)}$

(3) $f(t) = \frac{K_K}{1+a\exp(-bt)+\frac{b \cdot a_K}{b-b_K}\exp(-b_K t)}$

[b]Figures in parentheses indicate t-value.

telephones, bi-logistic growth demonstrates a better fit than the logistic growth function within a dynamic carrying capacity for color TV sets and personal computers.

(c) This implies that while color TV sets have provided viewers with new functionality these new functions are limited, and personal computers have been developing in a co-evolutionary way with workstations and/or traditional technology primarily with advanced type of word processors.

3.2.2.4 Functionality Development by Self-Propagating Behavior

These analyses demonstrate that innovative goods with the highest IT density as personal computers and cellular telephones match the logistic growth function within a dynamic carrying capacity. This suggests that the behavior of a dynamic carrying capacity has some relevance with IT's functionality development.

Equation (vi) in Sect. 3.2.2 (1) demonstrates that the dynamic carrying capacity $K(t)$ increases together with increase of the number of adopters (customers) $f(t)$ as time goes by. Increase in $K(t)$ induces $f(t)$, which in turn activates interactions with institutions leading to an increase in potential customers (carrying capacity) by increasing the value and function stimulated by network externalities. This dynamism can be depicted as a mechanism illustrated in Fig. 3.8. Thus, IT's specific features, or functionality, are formed in this interactive process.

Therefore, we postulate that IT creates new demand in this development process. Specific functionality is formed in this interactive self-propagating behavior.

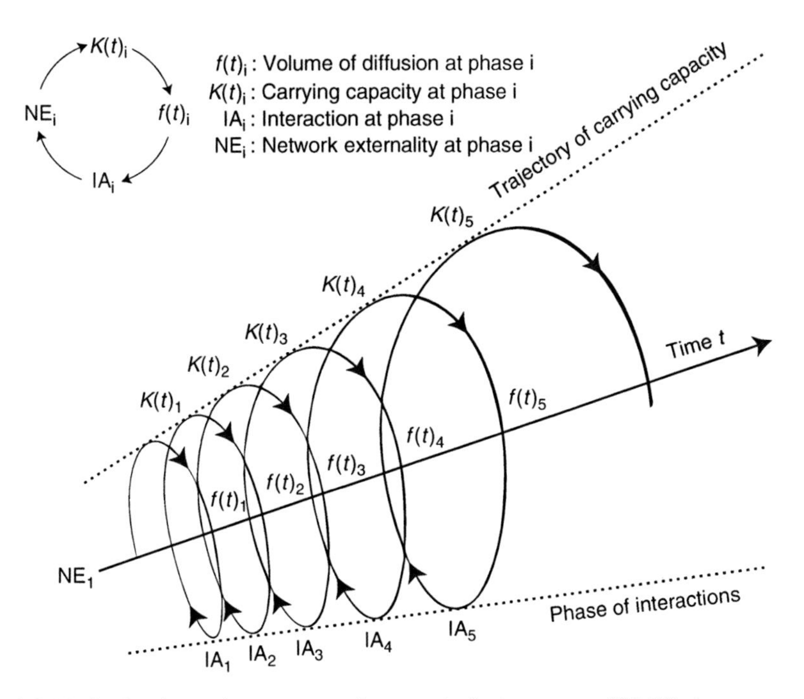

Fig. 3.8 Mechanism in creating a new carrying capacity in the process of IT diffusion

In order to demonstrate this postulate, an attempt was conducted to decompose contribution factors of TFP (total factor productivity) increase. Following Nadiri and Schankerman [22], and introducing technology stock as a source of the impact of direct and indirect technology change, TFP change was developed into the following equation:

$$\dot{\mathrm{TFP}} = (1-k^{-1}\eta)\dot{F}_{\mathrm{d}} - (1-k^{-1}\eta)\psi\eta\sum_i s_i\dot{P}_i + (1-k^{-1}\eta)\eta^2(\psi-1)k^{-1}\cdot \frac{\partial V}{\partial T}\cdot\frac{T}{V}\cdot\dot{T} + k^{-1}\eta\cdot\frac{\partial V}{\partial T}\cdot\frac{T}{V}\cdot\dot{T},$$

where V, GDP; F_{d}, final demand; T, technology stock; P, factor's price; $s_i = \frac{P_i X_i}{PV}$; X_i, factor i's quantity; η, production elasticity to cost; e, elasticity to production; $\psi = \frac{e}{1-e(1-\eta)}$; k, profit ratio $\left(=\frac{PV}{C}\right)$; and C, total cost.

The first term represents impacts of an exogenous shift of product demand, the second term represents impacts of change in factor prices, and the third and the fourth terms represent impacts of indirect and direct technology change, respectively.

Figure 3.9 illustrates the results of the analysis which demonstrates a significant correlation between IT intensity[6] and the contribution of the exogenous shift of product demand increase to TFP increase in major sectors of Japan's manufacturing industry over the period 1995–1998. This analysis supports the foregoing postulate that IT creates new demand which induces a TFP increase leading to a spiral development as illustrated in Fig. 3.10.

3.2.3 Features of IT

As examined in Sect. 3.2.2, personal computers and cellular telephones that contain the highest IT density are technologies that matched the logistic growth function within a dynamic carrying capacity. This is in line with the postulation that these technologies are self-propagating because of the nature of their interactivity. Consequently, IT's epidemic behavior closely interrelates with the continuous increase in the number of potential users. This means that during the course of diffusion, IT interacts with individuals, organizations, and society as a whole, dynamically transforming its functionality, and extending potential users in line with these newly acquired features. Furthermore, this characterizes the unique diffusion process of IT in that it alters the carrying capacity or creates a new carrying capacity in the process of its diffusion, thereby acquiring new specific features.

Figure 3.11 compares the diffusion process of manufacturing technology and IT. As Fig. 3.11 describes, IT's diffusion process is stimulated by an interaction with institutions and institutional change is also stimulated by an interaction with IT,

[6] See Appendix for measurement of IT intensity.

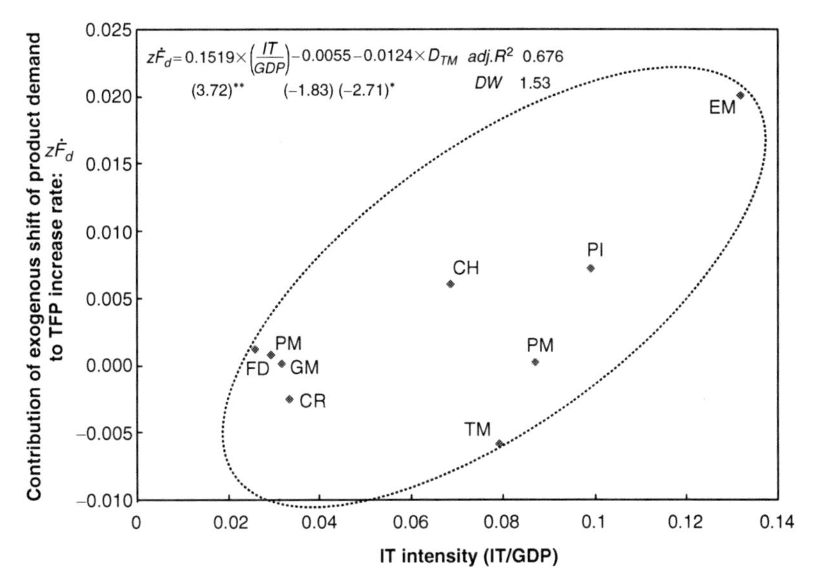

[a] $z \equiv 1 - k^{-1}\eta$ where k: profit ratio; and η: productionelasticity to cost.

[b] ** and * indicate statistically significant at 1% and 5%, respectively

[c] EM: electrical machinery; PI: precision instruments; CH: chemicals; PM: primary metals; TM: transportation equipment; GM: general machinery; CR: ceramics; PP: pulp and paper; and FD: food.

[d] D_{TM}: dummy variable

Fig. 3.9 Correlation between IT intensity and TFP increase through creation of final demand in Japan's manufacturing industry (1995–1998)

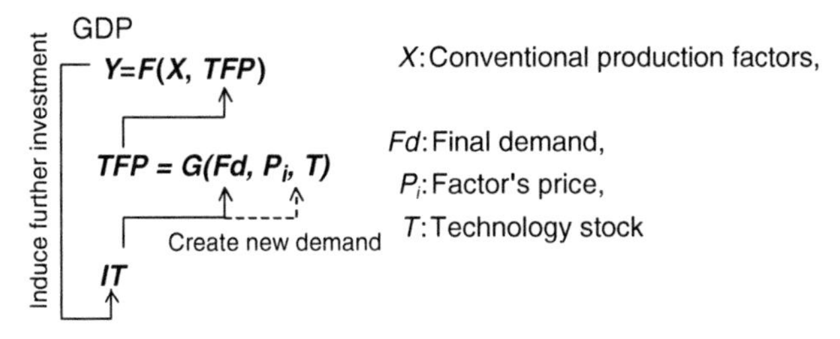

Fig. 3.10 Self-propagating structure of IT

leading to a co-evolution of the technology itself and institutions. In this process, rising technology value increases the number of potential users and a "virtuous cycle" results.

This behavior indicates that IT behaves differently because of some unique features facilitated by the institutions involved in the innovation process. Whether a national can fully exploit the benefits of IT greatly depends on the nation's institutional ability to flexibly respond and adopt this technology given its unique features. In other words "institutional elasticity" affects competitiveness [23]. We

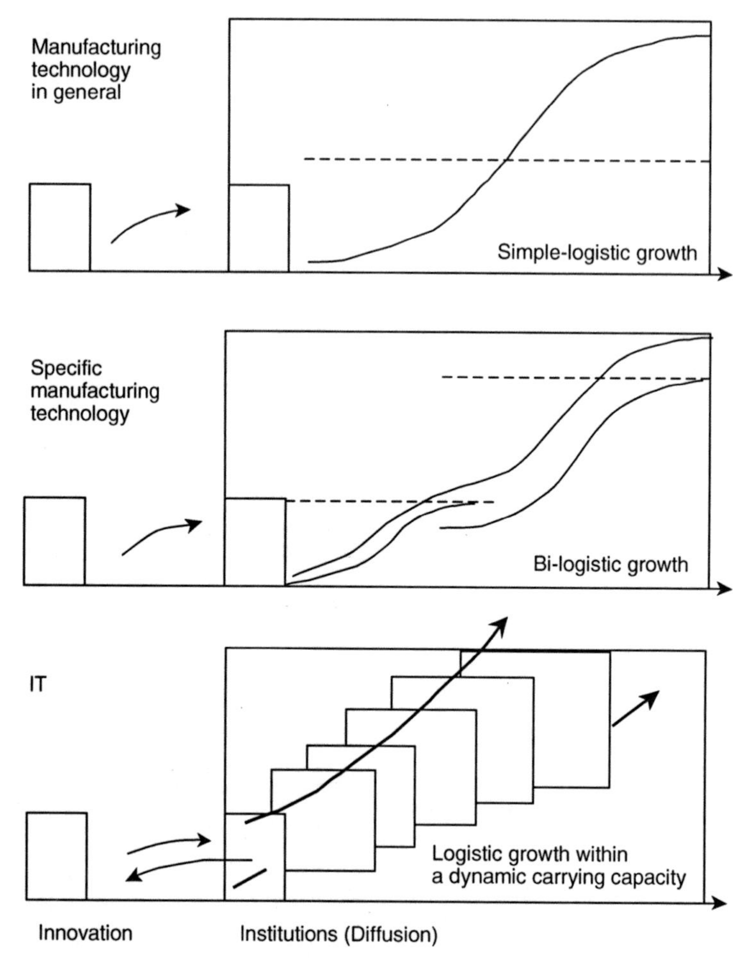

Fig. 3.11 General concept of technology diffusion process – comparison between manufacturing technologies and IT

have identified the following unique features of IT which must be compatible with firm strategy and development in order to enhance and facilitate competitiveness:

– *Disseminative*

This is represented by a logistic growth function within a dynamic carrying capacity. The dynamic evolution of a carrying capacity can be directly connected to a disseminative feature of IT. The rapid development of IT together with network externalities enable IT related products and services to disseminate rapidly. In this context, appropriately judging the surrounding environment and quickly commercializing new products and services are crucial to survive in an information society [24]. To make the best use of the disseminative nature of IT, organizations must make decisions quickly and react flexibly to changing environments. The increasing significance of global technology spillover is important in this context [25].

– *Interactive*

The leading player of the IT industry is now shifting from personal computers to networks [26]. Each time people use networks to communicate or exchange information, they actually interact with IT, and the phenomena of network externalities push up the carrying capacity raising the value of networks through the interaction between people and IT. Accordingly, the interactiveness of IT is an important feature explaining this unique behavior.

Inside organizations, the interactive nature of IT improves the efficiency of the decision making process and induces structural transformation of organizations from hierarchical to networked [2]. In order to exploit the potential benefits of IT, a reorganization of work that introduces new work practices is necessary [27].

– *Co-evolutional*

As Figs. 3.8 and 3.11 illustrate, features of IT and institutions evolve together during the course of their interaction. IT diffuses throughout the society as a type of social infrastructure and transforms economic, social, and cultural system of nations. In an information society, where IT functions as social infrastructure, growth depends more than ever on responding to changing demands of the workplace and society more broadly [27]. In this context, the most effective way to maximize the benefits of IT should be the evolution of the society itself as technological development proceeds.

Institutional elasticity enables these features to continually transform business practice [32].

3.3 Implications

If a nation's indigenous institutions can react flexibly to incorporate IT, the diffusion process of IT is accelerated. The nation can then the nation exploit the potential benefits of IT, resulting in enhanced international competitiveness. Thus, institutional elasticity and IT features construct a subtle cycle increasing a firms' ability to compete and thus creating a more competitive business environment. Japan enjoyed a "virtuous cycle" in terms of its competitive environment prior to the 1980s [25]. In this cycle, subtle relationships between technology, labor and capital matched competitive requirements in an industrial society [28]. However, under the current information society paradigm, this cycle collapsed. One of the reasons is that Japan's institutional system was flexible enough to enhance the growth behavior in manufacturing technology innovation is not elastic enough to fully reap the benefits of IT.

Although a detailed description of the Japanese institutional system is beyond the scope of this paper, the mechanism of Japan's competitive stagnation in the 1990s is briefly discussed below.

The fairly homogeneous population, together with a highly dense population in Japan, has contributed to the development of unique Japanese organizational and behavioral norms such as group orientation and feeling comfortable "being the same" as others. McMillan [29] concisely characterized the Japanese as consensual, highly stable, homogeneous, disciplined, and long-term oriented.

During the "catching-up" period prior to the 1990s, manufacturing technology was considered as a core technology of an industrial society. In this context, Japanese business management systems such as lifetime employment, the seniority system, lean production (e.g. TQC, JIT and Kaizen), the main bank system, and *keiretsu* well matched the nation's institutions and successfully established the feeling of "family ties" strengthening the highly efficient closed network among related entities. Actually, within these closed networks, implicit transaction rules among related entities and specific communication "language" were developed that after all excluded new comers outside these tight-nit "families."

A new paradigm emerged with the dramatic advancement of IT in the 1990s, Japan can no longer depend on well-tried, low-risk paths and other benefits available to a country undergoing "catching-up" [30]. Furthermore, Japanese indigenous characteristics such as homogeneity and preferring high stability do not allow firms to react flexibly in enhancing the disseminative, interactive and co-evolutional features of IT. Consequently, Japan's institutional system, which performed effectively in the 1980s, is not efficient any more in an information society. In fact, it often hinders exploiting the potential benefits of IT.

In contrast, as MacRae [31] argued, the melting pot of the US makes the nation a great generator of new ideas, cultivates a frontier spirit, and enhances flexibility, thus enhancing and capitalizing on the unique features of IT. These indigenous qualities in the US efficiently accelerated the structural changes of the US industry in the 1990s.

These structural changes in US firms led the nation to be successful in an information society paradigm.

Figure 3.12 suggests how Japanese institutional systems and those of the US performed this paradigm shift from an industrial society to an information society. It demonstrates that while the Japanese institutional system was effective in the paradigm of an industrial society, it has not been effective in the new paradigm of an information society.

3.4 Conclusion

The effective utilization of IT in an information society will differ greatly depending on structural differences in different countries. In particular, the elasticity or flexibility of institutions in the national innovation system is critical. Furthermore, the specific functionality of IT is developed through interactions with flexible institutional systems. This chapter attempts to show this process by focusing on the unique diffusion process, or epidemic behavior of IT.

	1980s	*1990s*
Paradigm	**Industrial society**	**Information society**
Core technology	Manufacturing technology	IT
Key features	**Given, Provided by suppliers**	**To be formed during the course of interaction with institutions**
Actors responsible for features formation	Individual firms/organizations	Institutions as a whole
System structure	Optimization	Standardization

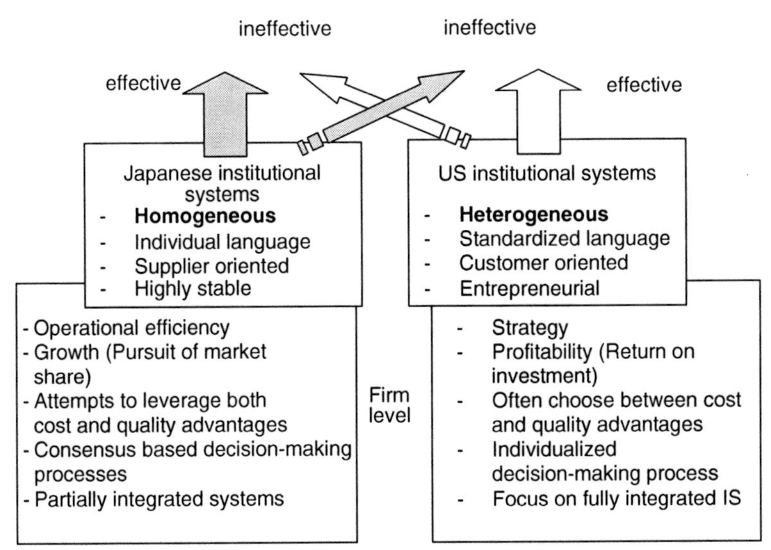

Fig. 3.12 Comparison of effectiveness between Japanese institutional systems and the US institutional systems under the paradigm shift

First, through the mathematical analysis of the diffusion process of innovative goods, the mechanism in creating a new carrying capacity in the process of IT diffusion was conceptualized. On the basis of this framework, an empirical analysis of Japan's manufacturing industry using a new approach to decomposing the factors of TFP was developed. This postulated that IT creates new demand in its development process and specific functionality is formed in this interactive system. Second, an empirical analysis of the diffusion process of innovative goods in Japan was conducted using refrigerators, fixed telephones, Japanese word processors, color TV sets, personal computers and cellular telephones which represent innovative goods across a spectrum from mono-function to multi-function products. Through a comparative analysis of epidemic behavior between these technologies using a logistic growth function within a dynamic carrying capacity, it was demonstrated that innovative goods with higher IT density matched the logistic growth function within a dynamic carrying capacity. These analyses demonstrated that the specific features of IT are formed through dynamic interaction with an institutional system. In addition, certain IT qualities enabling this self-propagating behavior were identified and discussed. These unique qualities critical to enhancing the epidemic behavior observed, depend on institutional flexibility to "capture" the potential of IT. Furthermore, Japan's institutional system currently does not provide the flexibility required to fully exploit these features.

These findings remind us of the significance of the role of institutional elasticity in making full utilization of IT and also the urgency of recreating institutional elasticity in Japan. While this paper briefly reviews this subject, substantial analysis is beyond the scope this work. Further analysis at the firm level in terms of fundamental approaches to information and process systems as well as strategic management is required to fully elucidate critical success factors in US and Japanese firms amidst the information society paradigm.

Appendix. Data Construction and Sources

Appendix 1. TFP and IT Intensity

In order to measure TFP and IT intensity, the following production function was used:

$$V = F\left(L, K, I, T, t\right),$$

where A, scale coefficient; L, labor; K, capital; I, IT production factor; I_l, IT labor; I_k, IT capital; T, technology stock; and t, time trend. Duplication among each production element was deducted.[7]

IT production factor was constructed using the data from the Ministry of International Trade and Industry's (MITI).[8] "Current Status of Japanese Information Processing," which referred the "Survey on Information Processing in Japan" by Japan Information Processing Development Center. "Capital Matrix of the Input–Output Tables" was also used to supplement the IT related investment that is not covered by the Survey. The resultant IT production factor is explained by the IT related investment listed in Table 3.4.

Appendix 2. Epidemic Behavior

In the analysis of the epidemic behavior of six innovative goods, trends in the cumulative number of adopters were analyzed using an epidemic function.

[7] See [30] for data construction and sources for L, K and T.

[8] MITI renamed the Ministry of Economy, Trade and Industry on January 6, 2001 under the structural reform of the Japanese government.

Table 3.4 IT related investment

Labor cost		Outsourced personal expenses, education and training cost, personal expenses, service charge, etc.
Capital cost	Hardware	Depreciation cost, rent fee, lease fee, installation charge, maintenance charge
	Software	Use charge, purchase cost, programming charge, consignment cost, machine rent charge, calculation consignment cost, data input charge
	Network	Network charge, network subscription charge, online service charge

Provided that the "pregnancy" period is short enough to neglect this timing and the depreciation rate can be treated as a reverse of the lifetime, the cumulative number is measured by the following equation:

$$N_t = P_t + (1-\rho)N_{t-1},$$
$$N_0 = \frac{P_1}{g+\rho},$$
$$\rho = \frac{1}{\mathrm{LT}},$$

where N_t, cumulative number of adopters at time t; P_t, number of shipment for domestic use at time t; g, increase rate of production in the initial period; ρ, depreciation rate; and LT, life time (average years in use).

A.2.1 Refrigerators (1951–1999)

The annual shipment volume of refrigerators from the year 1951 to 1999 was obtained from the "Report on Machinery Statistics" (Ministry of International Trade and Industry (MITI), annual issues). Since the ratio of imports and exports of refrigerators to their shipment as a whole has not been changing greatly, the annual shipment volume was used for the analysis.

The depreciation rate was measured by multiplying the rate of obsolescence of technology in the electrical machinery industry [25] and the ratio of depreciation rate of refrigerators ("Consumer Confidence Survey" (Cabinet Office, 1998–2001)) and electrical machinery industry in 1998.

A.2.2 Fixed Telephones (1953–1999)

The cumulative number of fixed telephones subscribers from 1953 to 1999 was obtained from NTT's annual reports.

A.2.3 Japanese Word Processors (1982–1997)

The annual domestic production volume of Japanese word processors from the year 1982 to 1997 was obtained from "Industry in Japan: A Graphical Look at 1626 Goods and Services" (Development Bank of Japan, Economic & Industrial Research Department, 1999).

The depreciation rate was measured by multiplying the rate of obsolescence of technology in the electrical machinery industry [25] and the ratio of depreciation rate of Japanese word processors ("Consumer Confidence Survey" (Cabinet Office, 1998–2001)) and the electrical machinery industry in 1998.

A.2.4 Color TV Sets (1966–2000)

The annual domestic shipment volume of color TV sets for domestic use from the year 1966 to 2000 was obtained from the "Survey of Japan Electronics and Information Technology Industries Association" (JEITA, annual issues).

The depreciation rate was measured by multiplying the rate of obsolescence of technology in electrical machinery industry [25] and the ratio of depreciation rate of color TV sets ("Consumer Confidence Survey". (Cabinet office government of Japan, 1998–2001)) and electrical machinery industry in 1998.

A.2.5 Personal Computers (1987–2000)

The 32-bit personal computer was analyzed as the PC. The quarterly shipment of the 32-bit PC for domestic use from 1987 to 2000 was obtained from the "Personal Computers Statistics" (Japan Electronics and Information Technology Industries Association (JEITA), annual issues).

The depreciation rate from 1998 to 2000 was estimated 20% p.a. by using the reverse of legal life time defined by the Corporate Tax law. The rate before 1998 was measured by multiplying the rate of obsolescence of technology in electrical machinery industry [25] and the ratio of depreciation rate of PC and electrical machinery industry in 1998.

A.2.6 Cellular Telephones (1996–2001)

The cumulative number of cellular telephones contracts from September 1996 to March 2001 (monthly statistics) was obtained from monthly reports issued by Telecommunications Carriers Association (TCA).

References

1. OECD, Special Issue on Information Infrastructures, STI Review (OECD, Paris, 1997)
2. Telecommunications Council, Japan, The Info-Communications Vision for the 21st Century (Telecommunications Council for the Minister of Posts and Telecommunications, 2000, Tokyo)
3. V.W. Ruttan, Technology, Growth, and Development – An Induced Innovation Perspective (Oxford University Press, New York, 2001)
4. US DOC, Digital Economy 2000 (DOC, Washington, DC, 2000)
5. F. Cairncross, The Death of Distance (Harvard Business School Press, Boston, 1997)
6. US DOC, Falling Through the Net: Toward Digital Inclusion (DOC, Washington, DC, 2000)
7. H. Binswanger and V. Ruttan, Induced Innovation: Technology, Institutions, and Development (John Hopkins University Press, Baltimore, 1978)
8. E.M. Rogers, Diffusion of Innovations (The Free Press of Glencoe, New York, 1962)
9. Z. Griliches, Hybrid Corn: An Explanation in the Economics of Technical Change, Econometrica 25, No. 4 (1957) 501–522
10. P.S. Meyer, Bi-logistic growth, Technological Forecasting and Social Change 47, No. 1 (1994) 89–102
11. E. Mansfield, Intrafirm Rates of Diffusion of an Innovation, The Review of Economics and Statistics 45, No. 4 (1963) 348–359
12. E. Mansfield, Industrial Research and Technological Innovation: An Econometric Analysis (Longman, London, 1969)
13. J.S. Metcalfe, The diffusion of innovation in the lancashire textile industry, Manchester School of Economics and Social Studies 2, 1970, 145–162
14. K. Norris and J. Vaizey, The Economics of Research and Technology (George Allen & Unwin, London, 1973)
15. C. Marchetti, On strategies and fate, in Hafele et al. (ed.), Second Status Report on the IIASA Project on Energy Systems 1975 (IIASA, Laxenburg, Austria, 1976) 203–217
16. C. Marchetti and N. Nakicenovic, The Dynamics of Energy Systems and the Logistic Substitution Model, IIASA Research Report RR-79-13 (IIASA, Laxenburg, Austria, 1979)
17. R. Coombs, P. Saviotti and V. Walsh, Economics and Technological Change (Macmillan, London, 1987)
18. S.M. Oster, Modern Competitive Analysis (Oxford University Press, New York, 1994)
19. T.C. Schelling, Social mechanisms and social dynamics, in P. Hedstrom and R. Swedberg (eds.), Social Mechanisms: An Analytical Approach to Social Theory (Cambridge University Press, Cambridge, 1998) 32–44
20. P.S. Meyer and J.H. Ausbel, Carrying capacity: a model with logistically varying limits, Technological Forecasting and Social Change 61, No. 3 (1999) 209–214
21. Ministry of Posts and Telecommunications (MPT), Japan, White Paper 2000 on Communications in Japan (MPT, Tokyo, 2000)
22. M.A. Nadiri and M.A. Schankerman, The Structure of Production, Technological Change, and the Rate of Growth of Total Factor Productivity in the U.S. Bell System, in Productivity Measurement in Regulated Industries (Academic Press, Inc., New York, 1981) 219–247
23. C. Watanabe and R. Kondo, Institutional Elasticity towards IT Waves for Japan's Survival – The Significant Role of an IT Testbet, Technovation 23, No. 3 (2003) 205–219
24. Economic Planning Agency, White Paper on the Japanese Economy (Tokyo, 2000)
25. C. Watanabe, Systems Option for Sustainable Development, Research Policy 28, No. 7 (1999) 719–749
26. D.C. Moschella, Waves of Power (AMACOM, New York, 1997)
27. OECD, The New Economy: Beyond the Hype, Final Report on the OECD Growth Project (OECD, Paris, 2001)
28. C. Watanabe, The Feedback Loop between Technology and Economic Development: An Examination of Japanese Industry, Technological Forecasting and Social Change 49, No. 2 (1995) 127–145

29. C. McMillan, The Japanese Industrial System (Walter de Gruyter & Co., Paris, 1996)
30. R. Aggarwal, The shape of post-bubble japanese business: preparing for growth in the new millennium, in International Executive, 38, No. 1 (Wiley, New York, 1996) 9–32
31. H. MacRae, The World in 2020: Power, Culture and Prosperity (Harvard Business School Press, Boston, 1995)
32. C. Watanabe, B. Zhu, C. Griffy-Brown and B. Asgari, Global Technology Spillover and Its Impact on Industry's R&S Strategies, Technovation 21, No. 5 (2001) 281–291

Chapter 4
A Substitution Orbit Model of Competitive Innovations

Abstract Successful innovation and diffusion of technology can be attributed to the identification of the orbit of emerging new technologies that complement or substitute for existing technologies. This dynamism resembles the co-evolution process in an ecosystem. In an ecosystem, in order to maintain sustainable development, the complex interplay between competition and cooperation, typically observed in predator–prey systems, create a sophisticated balance. Given that an ecosystem can be used as a masterpiece system, this sophisticated balance can provide suggestive ideas for identifying an optimal orbit of competitive innovations with complement or substitution dynamism.

Prompted by such a sophisticated balance in an ecosystem, this chapter analyzes the optimal orbit of competitive innovations, and on the basis of an application of Lotka–Volterra equations, it reviews substitution orbits of Japan's monochrome to color TV system, fixed telephones to cellular telephones, cellular telephones to mobile internet access service, and analog to digital TV broadcasting. On the basis of substitution orbits analyses, it attempts to extract suggestions supportive to identifying an optimal policy option in a complex orbit leading to expected orbit.

Key findings include policy options that are effective in controlling parameters for Lotka–Volterra equations leading to expected orbit.

Reprinted from *Technological Forecasting and Social Change* 71, No. 4, C. Watanabe, R. Kondo, N. Ouchi and H. Wei, A Substitution Orbit Model of Competitive Innovations, pages: 365–390, copyright (2004), with permission from Elsevier.

4.1 Introduction

Given the global paradigm shift from an industrial society to an information society, optimal control of diffusion orbit of competitive innovations would be crucial for nation's competitiveness [1]. While advanced technologies that substitute for old technologies are usually welcomed, it is not always a simple matter to gain general

C. Watanabe, *Managing Innovation in Japan: The Role Institutions Play in Helping or Hindering how Companies Develop Technology*, DOI: 10.1007/978-3-540-89272-4_4,

consensus for the transition of such technologies [2]. Crucial issues related to the potential benefits of innovation can thus be focused on as innovation in transition [3].

"For most companies today, the only truly sustainable advantage comes from out-innovating the competition. Successful businesses are those that evolve rapidly and effectively. Yet innovative business can't evolve in a vacuum" [4]. Indeed, socio-economic development can mainly be attributed to innovation evolved from the institutions [5–7] leading to advanced technologies that substitute for (sometimes complement with) existing technologies.

In tracing innovation diffusion, it is generally considered the spread of exchangeable mutually exclusive innovations from the ecological viewpoint, as competition between the innovations within an active social and physical environment which can be classified into the following four categories [8]:

(a) The interaction between innovations themselves on the basis of competition,
(b) The interaction between adopters of innovations on the basis of exchange of information about the utility of innovations for adopters,
(c) The individual's dynamic choice process, and
(d) The interaction of an active social and physical environment in the interaction between individuals, in the choice process, and in the competition between alternative innovations.

Sonis [8] stressed that it is important to underline that the social and physical environment is active if it changes the behavior of alternatives and individuals: (a) implicitly by filtering and directing or intensifying the information flows between the individuals and between the individuals and alternatives, and (b) explicitly by physical, social cultural, etc. restrictions and prohibitions, or by support and stimulation.

These postulates stimulate the significance of an ecosystem approach in identifying the substitution orbit of competitive innovations. Moore [4] complains that, contrary to an increasing significance, current business strategies such as networks, under the rubric of strategic alliances, virtual organizations, and the like provide little systematic assistance for managers who seek to understand the underlying strategic logic of change. He stressed the significance of a systematic approach to strategy with a view not as a member of a single industry but as part of a business ecosystem that crosses a variety of industries. In a business ecosystem, according to Moore, companies co-evolve[1] capabilities around a new innovation: they work cooperatively and competitively to support new products, satisfy customers needs, and eventually incorporate the next round of innovations. Thus, focus of business priority, business ecosystem should be the process of co-evolution: the complex interplay between competitive and cooperative business strategies.

Lotka–Volterra equations analyze the behavior of two (or more) interacting species within a limited environment. As a special case they also analyze the such complex interplay between competitive and cooperative game in an ecosystem, particularly predator–prey systems [9]. Lotka–Volterra principle for this case can be

[1] Moore refers Gergory Rateson's definition of co-evolution that as a process in which interdependent species evolve in an endless reciprocal cycle – in which "changes in species A set the stage for the natural selection of changes in species B" – and vice versa.

summarized as "two closely similar species will not both indefinitely be able to occupy essentially the same ecological niche, but that *the slightly more successful of two* will completely supplant the other eventually" [10]. This principle represents the foregoing dynamic game in the process of co-evolution in an ecosystem. Therefore, Lotka–Volterra equations which can be applied also to general logistic growth in a single innovation are useful for the analysis of a substitution orbit of competitive innovations with complex interplay.

Prompted by such postulate, this paper analyzes the substitution orbit of competitive innovations, and on the basis of an application of Lotka–Volterra equations, it reviews substitution orbits of Japan's monochrome to color TV system, fixed telephones to cellular telephones, cellular telephones to mobile internet access service, and analog to digital TV broadcasting. Some substitution orbits of these innovations can be traced as the diffusion orbit of a single technology, not as a part of competing or cooperating system, by applying logistic growth function which is a simplified structure of Lotka–Volterra equations. However, it is insufficient in case of the competitive systems with complex interplay in such a shift from analog to digital TV broadcasting as it requires the complex orbit conditions including rapid shift from analog to digital TV broadcasting system and simultaneous service by both systems. An attempt to extract suggestions supportive to identifying an optimal policy option in such a complex orbit.

Section 4.2 reviews substitution orbits for transition of competitive innovations. Section 4.3 outlines model synthesis for two-dimensional Lotka–Volterra equations. Section 4.4 attempts to extract suggestions for policy option in a complex orbit. Section 4.5 briefly summarizes implications for the substitution orbit of competitive innovations.

4.2 Substitution Orbit of Competitive Innovations

4.2.1 Substitution Orbit: Japan's Experiences

4.2.1.1 Substitution from Monochrome to Color TV System

Figure 4.1 depicts the diffusion orbits of TV sets in Japan. Television broadcasting was inaugurated in Japan in 1953, followed by the color TV broadcasting in 1960. As the figure shows, the diffusion of color TV sets rapidly increased and the diffusion level of color TV sets to Japanese households exceeded that of monochrome TV sets around 1973 when color TV broadcasting became available in all TV programs.

Since the technical standard for the color TV broadcasting was compatible to monochrome TV sets, people could receive color TV broadcasting by their monochrome TV sets as monochrome TV programs. In this sense, color TV broadcasting and monochrome TV broadcasting were in a competitive relationship because people could choose either monochrome or color TV sets according to their

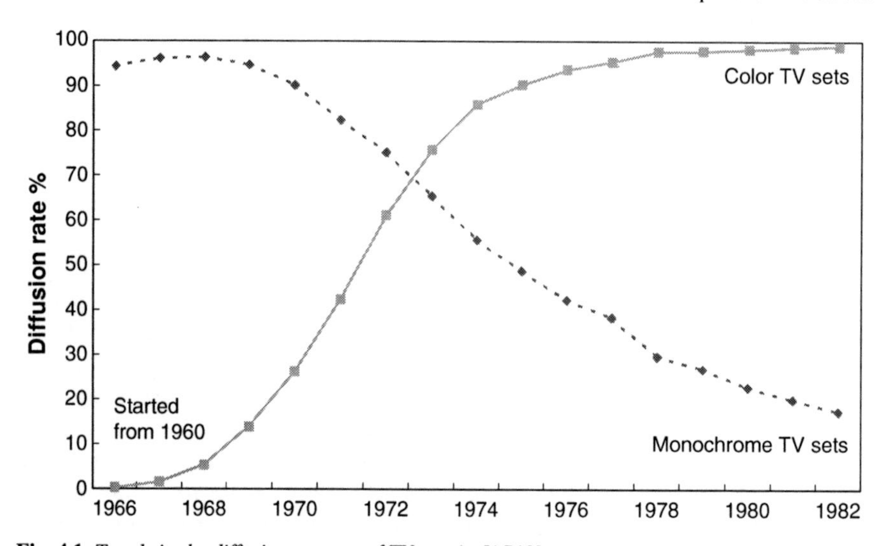

Fig. 4.1 *Trends in the diffusion process of TV sets in JAPAN*
[a] Diffusion level is represented by diffusion ratio (ratio of holders and households: %)
Source: "Consumer confidence survey," Cabinet Office, Government of Japan

preferences. As the price of color TV sets declined, people started to switch from monochrome to color TV sets because the function of color TV sets are clearly superior to monochrome TV sets. Consequently, color TV sets diffused rapidly in line with a logistic growth. They exhibited rapid growth from 1968, 8 years after their inauguration by substituting for monochrome TV sets.

4.2.1.2 Substitution from Fixed Telephones to Cellular Telephones

Figure 4.2 illustrates the trends in the diffusion process of fixed telephones and their transition to cellular telephones in Japan. As the figure shows, the number of subscribers exhibited a rapid increase from the middle of the 1960s lasting two decades up to the middle of the 1980s. However, during the 1990s, it gradually stagnated and started to decrease from 1996.

The decrease or stagnation in the number of subscribers to the fixed telephone can be partly attributed to the emergence of the cellular telephones, which dramatically developed their subscribers in the 1990s triggered by deregulations such as abolishment of a deposit in 1993 and the liberalization of the sales of terminal receivers in 1994.

Figure 4.2 also compares the transition of the number of subscribers to fixed telephones as well as cellular telephones from the year 1992 to 2000 in order to visualize the impact of cellular telephones to fixed telephones. At the end of the fiscal 2000, the number of subscribers to the cellular telephones exceeded that of fixed telephones.

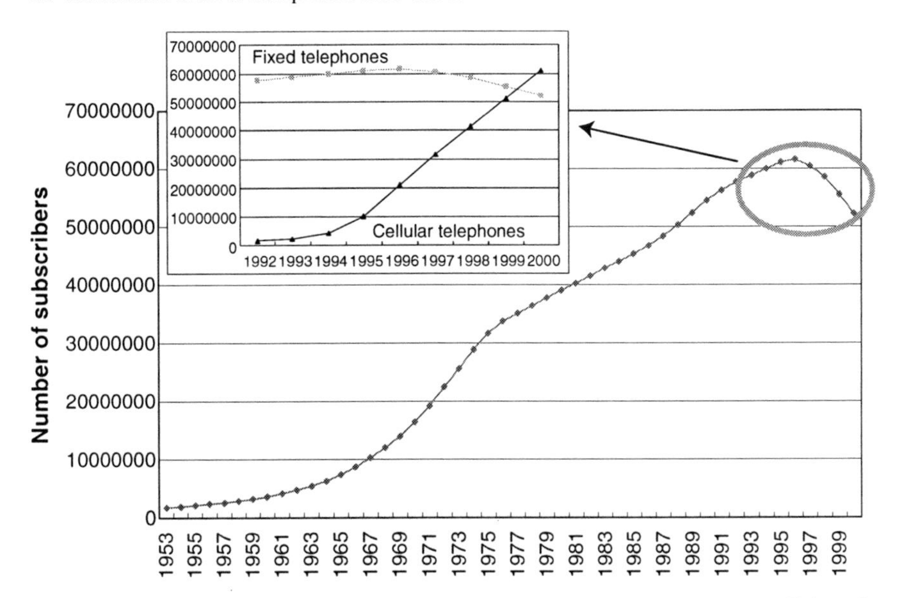

Fig. 4.2 *Trends in the diffusion process of fixed telephones and their transition to cellular telephones in Japan (1953–2000)*
[a] The part in the right-down corner of this figure illustrates transition of the number of subscribers to fixed telephones and cellular telephones

Cellular telephones also followed a logistic growth and could achieve very rapid diffusion, 4 years after their start of substantial service because when the service became available at a reasonable price, people already well knew what the telephone is and regarded them as a complementary method to fixed telephones which cannot be used outside. Recently, some people even possess only cellular telephones because basically cellular telephones can substitute for the functionality of fixed telephones.

In this sense, the relationship between fixed telephones and cellular telephones can be considered as both complementary and substitute.

4.2.1.3 Substitution from the Cellular Telephones to Mobile Internet Access

Figure 4.3 illustrates the diffusion process of cellular telephones with and without the mobile internet access service. The pioneer in this mobile internet access service that enables users to access the internet from their cellular handsets was NTT DoCoMo's *i-mode* service. While the mobile internet access also follows a logistic growth, since NTT DoCoMo introduced *i-mode* service in February 1999, this kind of mobile internet access service has been dramatically expanding and the number of subscribers reached 31.4 million as of February 2001.

This exceptionally rapid diffusion of the mobile internet access service can be explained by such factors as (a) cellular telephones were already diffused, that is, about 40.5 million users already existed in the market when *i-mode* service was

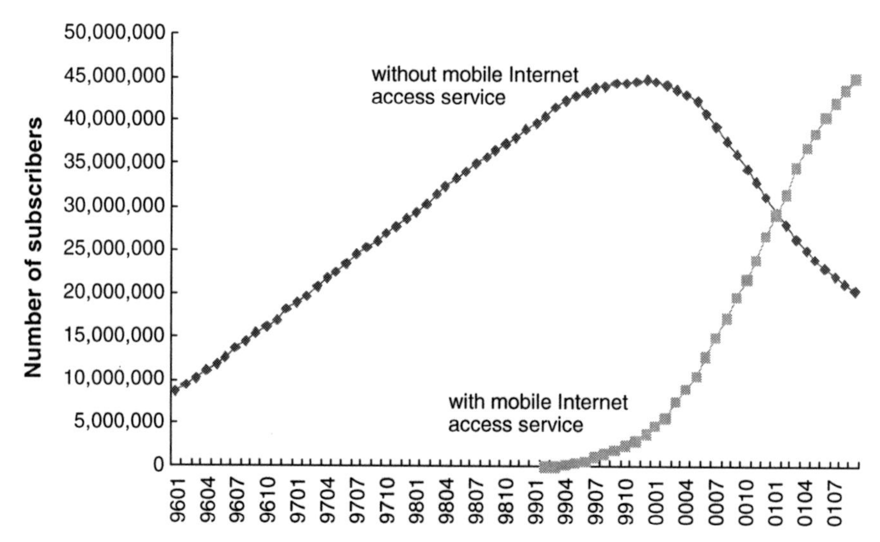

Fig. 4.3 *Trends in the diffusion process of the cellular telephone with and without the mobile internet access service (1996–2001)*
Source: Ministry of Public Management, Home Affairs, Posts and Telecommunications (MPHPT), http://www.tca.or.jp/

inaugurated, and (b) potential users somewhat recognized how useful the internet is because the internet was getting popular in Japan.

4.2.1.4 Substitution from Analog to Digital TV Broadcasting System

With the recent development of digital technology, that enabled effective error correcting, efficient data compression, and manageability of data, the TV broadcasting industry is facing a radical transitional phase – from analogue to digital system [11].

The digital TV broadcasting is highly expected to realize such advanced services as a variety of information services by data casting, interactive services which allows viewers to participate in TV programs, less deterioration in the quality of screen images, and manageable closed captions [12, 13].

Realizing the potential of the digital TV broadcasting, the substitution from analog to digital broadcast system has been proceeding worldwide as illustrated in Fig. 4.4.

The figure demonstrates that in the US, the digital satellite broadcast started in 1994, digital cable broadcast in 1997, and digital terrestrial broadcast in 1998. As for Japan, digital satellite broadcasting started in 1996 and cable TV in 1999. However, the digital terrestrial broadcasting is scheduled to be started in 2003 while Western Europe was somewhat quicker than Japan to move away from its previous arrangements [14].

Though the delay can be partly excused by the existence of excellent analog HDTV technology developed by NHK (the Japan Broadcasting Corporation), it is

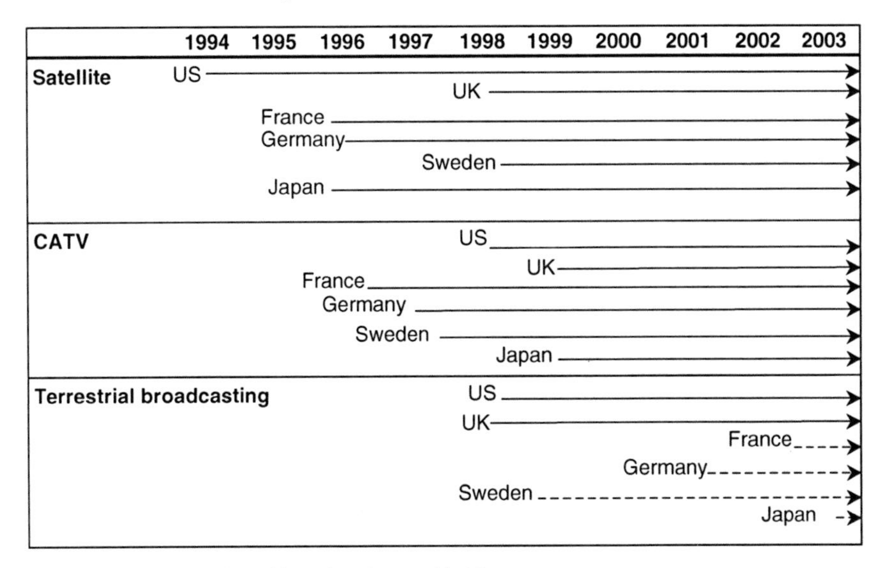

Fig. 4.4 *Trends in the digital broadcasting worldwide*
[a] *Dotted lines* imply that the service is expected to start
Source: White Paper 2001 on Communications in Japan, MPHPT [20]

actually a great concern because the digital terrestrial broadcasting is expected to have a significant positive impact both on society and economy as TV broadcasting is one of the most familiar media with almost a 100% diffusion ratio to Japanese households. According to the report by the Advisory Committee on Digital Terrestrial Broadcasting issued in October 1998 [15], introduction of the digital terrestrial broadcast is forecasted to create about 212 trillion yen (1.9 trillion US $) economic effect and about 7 million employment in 10 years, expecting the emergence of new services using the digital broadcast characteristics.

However, the transition from analog to digital broadcasting involves huge efforts in Japan. First of all, frequency usage is fairly congested in Japan and changes in existing analog channels are inevitable in order to allot frequencies to the digital terrestrial broadcasting. This *analog channel change* considerably affects viewers and broadcasters because the channel change requires adjustments of both receivers and transmitters.

In addition, a lack of information about new technology, fear of substitution and a reluctance to pay the cost of switching to new technology generally result in disturbing smooth transition. This is particularly the case with respect to Japan's switch from analog to digital TV broadcasting because of the high popularity of television to daily life of the Japanese people.

Under the dramatic advancement of IT, while hasty transfer sometimes accomplishes nothing, delayed transfer can result in a loss of national competitiveness. Thus, policy options for the optimal shift from the analog to digital TV broadcasting have become a crucial issue for Japan.

Thus, contrary to orbits of other transition examined, the orbit for Japan's transition from analog to digital terrestrial TV broadcasting is a unique and complex one as it should satisfy the following conditions:

(a) In order to minimize the impact of a transition delay, rapid shift from analog to digital TV broadcasting is strongly expected,
(b) While, as the US's experience in such shift advises, it is generally anticipated that shift from analog to digital in its initial stage is not necessarily easy,
(c) Simultaneous service by both analog and digital broadcasting should be provided over the period between the start of the digital terrestrial TV broadcasting service and the termination of the analog terrestrial TV broadcasting.

4.2.2 Comparative Assessment of Substitution Orbits

Based on the above comparative analysis, Table 4.1 summarizes the substitution orbits of competitive innovations: from monochrome to color TV, from fixed to cellular telephones, from cellular telephones to mobile internet access service, and from analog to digital TV broadcasting. Looking at the table we note that, among

Table 4.1 Comparison of substitution orbit for transition of competitive innovations

	Complement/ substitution	Years take for rapid growth	Sources of rapid growth	Government intervention for the transition
Monochrome TV/ color TV	Substitution	Eight years (logistic growth)	New function, virtuous cycle leading to cost reduction	Not substantial
Fixed telephones/ cellular telephones	Complement, partly substitution	Five years (logistic growth)	Less resistance to the new technology because of the familiarity to the existing similar technology	Deregulation
Cellular telephone/ Mobile internet access	Complement	One year (logistic growth)	New function, less resistance to the new technology because of the familiarity to the existing similar technology	
Analog TV/digital TV	Complementally substitution	?	New function, virtuous cycle leading to cost reduction	Substantial (vision, regulations, investment)

substitution orbits examined, years take for rapid growth have dramatically shortened as new functions increased.

As reviewed in Sect. 4.1, Lotka–Volterra equations, as a special case of Lotka–Volterra systems, analyzes such the complex interplay between competitive and cooperative species in an ecosystem by solving the following differential equations:

$$\begin{aligned} \dot{x} &= x(a - bx - cy) \qquad \dot{x} = \frac{\mathrm{d}x}{\mathrm{d}t}, \\ \dot{y} &= y(d - ex - fy) \end{aligned} \tag{4.1'}$$

where x, y, species in competitive or cooperative game; and a, b, c, d, e, f, positive coefficients.

This analysis provides a supportive suggestion to the substitution orbit of competitive innovation.[2]

Provided that x is preceding technology and y is new succeeding technology and given that y is strong enough than x, diffusion orbit of y can be approximated by a simple logistic growth as follows:

$$\dot{y} \approx y(d - fy). \tag{4.1''}$$

As reviewed in Figs. 4.1–4.3, substitution orbits of monochrome TV sets to color TV sets, fixed telephones to cellular telephones, and cellular telephones to mobile internet access service can be traced by a logistic growth within a single innovation as can be enumerated by (4.1)".

However, in case of such transition from analog to digital TV broadcasting system since this transition encompasses certain stable coexistence (simultaneous stage), both x and y demonstrate complex interplay. This is the case of Japan's transition from analog to digital TV system, Lotka–Volterra equations are useful for analyzing this interplay by identifying controlling parameters for expected orbit.

4.3 Lotka–Volterra Equations for Predator–Prey Systems

4.3.1 General Orbit Within Lotka–Volterra Equations

Given the rate of growth of species decreases linearly as a function of the density of species, interaction of two competing species x and y can be expressed by the following Lotka–Volterra equations:

$$\begin{aligned} \dot{x} &= x(a - bx - cy) = ax\left(1 - \frac{x}{a/b} - \frac{cd}{af}\cdot\frac{y}{d/f}\right) \equiv ax\left(1 - \frac{x}{k_x} - \alpha_{xy}\cdot\frac{y}{k_y}\right), \\ \dot{y} &= y(d - ex - fy) = dy\left(1 - \frac{ae}{bd}\cdot\frac{x}{a/b} - \frac{y}{d/f}\right) \equiv dy\left(1 - \alpha_{yx}\cdot\frac{x}{k_x} - \frac{y}{k_y}\right), \end{aligned} \tag{4.1}$$

[2] See flexible substitution models and the distinction between internal and external influence.

where $k_x(= a/b)$ and $k_y(= d/f)$ are carrying capacities, $\alpha_{xy}(= \frac{cd}{af})$ and $(= \alpha_{yx}$ $(= \frac{ae}{bd})$ are interaction coefficients, and a, b, c, d, e, f are positive coefficients (a and d, maximum diffusion scale).

Given that the orbit of x and y can be depicted by vector $V(x, y)$, and

$$V(x,y) = eH(x) + cG(y), \tag{4.2}$$

$$H(x) = \bar{x} \log x - x, G(y) = \bar{y} \log y - y, \tag{4.3}$$

$$\frac{1}{t_n}\int_0^{t_n} x(t)\mathrm{d}t = \bar{x}, \ \frac{1}{t_n}\int_0^{t_n} y(t)\mathrm{d}t = \bar{y}, \tag{4.4}$$

where $\bar{x}$ and $\bar{y}$, time average of x and y; and t_n, the period of the solution.

From $\frac{\mathrm{d}}{\mathrm{d}t}(\log x) = \frac{\dot{x}}{x} = a - bx - cy$

It follows by integration that $\int_0^{t_n} \frac{\mathrm{d}}{\mathrm{d}t} \log x(t)\mathrm{d}t = \int_0^{t_n} (a - bx(t) - cy(t))\mathrm{d}t$

i.e.

$$\log x(t_n) - \log x(0) = at_n - b\int_0^{t_n} x(t)\mathrm{d}t - c\int_0^{t_n} y(t)\mathrm{d}t.$$

Since $x(t_n) = x(0)$,

$$a = b\frac{1}{t_n}\int_0^{t_n} x(t)\mathrm{d}t + c\frac{1}{t_n}\int_0^{t_n} y(t)\mathrm{d}t = b\bar{x} + c\bar{y}. \tag{4.5}$$

Similarly,

$$d = e\bar{x} + f\bar{y}, \tag{4.5'}$$

while $\bar{x} = \frac{af-cd}{bf-ce}$ and $\bar{y} = \frac{bd-ae}{bf-ce}$.

Thus, orbit of x and y can be represented by (4.2), (4.3), (4.5) and (4.5)'.

The derivative of the function $V(x(t), y(t))$ by time t yields

$$\begin{aligned}\dot{V}(x,y) &= \frac{\partial V}{\partial x}\dot{x} + \frac{\partial V}{\partial y}\dot{y} \\ &= e\dot{H}(x)\dot{x} + c\dot{G}(y)\dot{y} \\ &= e\left(\frac{\bar{x}}{x} - 1\right)\{x(a - bx - cy)\} + c\left(\frac{\bar{y}}{y} - 1\right)\{y(d - ex - fy)\}.\end{aligned} \tag{4.6}$$

From (4.5) and (4.5)' we may replace a and d by $b\bar{x} + c\bar{y}$ and $e\bar{x} + f\bar{y}$, respectively. This yields

$$\begin{aligned}\dot{V}(x,y) &= e(\bar{x} - x)(b\bar{x} + c\bar{y} - bx - cy) + c(\bar{y} - y)(e\bar{x} + f\bar{y} - ex - fy) \\ &= be(x - \bar{x})^2 + 2ce(x - \bar{x})(y - \bar{y}) + cf(y - \bar{y})^2.\end{aligned} \tag{4.7}$$

By converting coordinates from (x, y) to (X, Y) where $X = x - \bar{x}$ and $Y = y - \bar{y}$, and twisting the axis of converted coordinates to (X', Y'), (4.7) can be developed to an elliptical orbit (in case when $bf > ce$) or a hyperbola (in case when $ce > bf$) as expressed in (4.8) and (4.9), respectively by using twisted coordinates (X', Y').[3]

[3] See the mathematical details and qualitative analysis of nonlinear systems by Lotka–Volterra approach.

$$\frac{X'^2}{\left(\sqrt{\frac{\dot{V}(x,y)}{\lambda_1}}\right)^2}+\frac{Y'^2}{\left(\sqrt{\frac{\dot{V}(x,y)}{\lambda_2}}\right)^2}=1 \text{ when } bf>ce, \tag{4.8}$$

$$\frac{X'^2}{\left(\sqrt{\frac{\dot{V}(x,y)}{\lambda_1}}\right)^2}-\frac{Y'^2}{\left(\sqrt{\frac{\dot{V}(x,y)}{-\lambda_2}}\right)^2}=1 \text{ when } ce>bf, \tag{4.9}$$

where

$$\lambda_1=\frac{be+cf+\sqrt{(be-cf)^2+4c^2e^2}}{2}. \tag{4.10}$$

$$\lambda_2=\frac{be+cf-\sqrt{(be-cf)^2+4c^2e^2}}{2}. \tag{4.11}$$

In order to identify twisted coordinates (X',Y'), the twisted angle θ should be measured which is expressed by the following equations (see Sect. 4.3.2):

$$\cos\theta=\frac{1}{\sqrt{1+\frac{\omega^2+2c^2e^2-\omega\sqrt{\omega^2+4c^2e^2}}{2c^2e^2}}}, \tag{4.12}$$

$$\sin\theta=\frac{1}{\sqrt{1+\frac{2c^2e^2}{\omega^2+2c^2e^2-\omega\sqrt{\omega^2+4c^2e^2}}}}, \tag{4.13}$$

where $\omega=be-cf$.

4.3.2 Factors Governing the Twisting of an Orbit

Provided that coefficients matrix of (4.7) **A** can be depicted as follows:

$$\mathbf{A}=\begin{bmatrix} be & ce \\ ce & cf \end{bmatrix}. \tag{4.14}$$

By using **A**, (4.7) can be depicted as follows:

$$\dot{V}(x,y)={}^t\mathbf{x}\mathbf{A}\mathbf{x}\ \mathbf{x}=\begin{bmatrix} X \\ Y \end{bmatrix}=\begin{bmatrix} x-\bar{x} \\ y-\bar{y} \end{bmatrix} \tag{4.15}$$

In identify factors governing the twisting angle θ, first, calculating determinants of matrix **A**:

$$|x\mathbf{E}-\mathbf{A}|=\begin{bmatrix} x-be & -ce \\ -ce & x-cf \end{bmatrix}=(x-be)(x-cf)-c^2e^2=0 \tag{4.16}$$

leads to the following equation:

$$x^2 - (be + cf)x + bcef - c^2e^2 = 0. \tag{4.17}$$

Solving (4.17), results in two eigen values for determinant of matrix **A** as follows:

$$\begin{aligned} \lambda_1 &= \frac{be + cf + \sqrt{(be + cf)^2 - 4bcef + 4c^2e^2}}{2}, \\ \lambda_2 &= \frac{be + cf - \sqrt{(be + cf)^2 - 4bcef + 4c^2e^2}}{2} \end{aligned} \tag{4.18}$$

The next step is to calculate the eigen vectors [$\mathbf{v}_1$ and $\mathbf{v}_2$] for matrix **A**. Since eigen values λ_1 and λ_2 are already known, two different cases can be considered:

(a) If $\lambda = \lambda_1$, then $\mathbf{A}\mathbf{v}_1 = \lambda_1 \mathbf{v}_1$.

$$\begin{bmatrix} be & ce \\ ce & cf \end{bmatrix} \begin{bmatrix} X_1 \\ Y_1 \end{bmatrix} = \lambda_1 \begin{bmatrix} X_1 \\ Y_1 \end{bmatrix}, \tag{4.19}$$

$$\begin{cases} beX_1 + ceY_1 = \lambda_1 X_1 \\ ceX_1 + cfY_1 = \lambda_1 Y_1 \end{cases}, \tag{4.20}$$

$$\begin{cases} (be - \lambda_1)X_1 + ceY_1 = 0 \\ ceX_1 + (cf - \lambda_1)Y_1 = 0 \end{cases}, \tag{4.21}$$

$$\mathbf{v}_1 = \begin{bmatrix} X_1 \\ Y_1 \end{bmatrix} = k_1 \begin{bmatrix} ce \\ \lambda_1 - be \end{bmatrix} (k_1 \neq 0). \tag{4.22}$$

(b) If $\lambda = \lambda_2$, then $\mathbf{A}\mathbf{v}_1 = \lambda_2 \mathbf{v}_1$.

Following the same simple algebraic steps as in case (a):

$$\mathbf{v}_2 = \begin{bmatrix} X_1 \\ Y_2 \end{bmatrix} = k_2 \begin{bmatrix} ce \\ \lambda_2 - be \end{bmatrix} (k_2 \neq 0). \tag{4.23}$$

The above eigen vectors $\mathbf{v}_1$ and $\mathbf{v}_2$ are linearly independent and they are perpendicular vectors forming a right angle ($k_1 \neq 0$ and $k_2 \neq 0$).

$$\mathbf{v}_1 = \begin{bmatrix} X_1 \\ Y_1 \end{bmatrix} = k_1 \begin{bmatrix} ce \\ \lambda_1 - be \end{bmatrix}. \tag{4.24}$$

The magnitude of vector $\mathbf{v}_1$ is calculated as follows:

$$\|\mathbf{v}_1\| = \sqrt{(cek_1)^2 + \{(\lambda_1 - be)k_1\}^2} = \sqrt{\left\{c^2e^2 + (\lambda_1 - be)^2\right\} k_1{}^2}. \tag{4.25}$$

If $\|\mathbf{v}_1\| = 1$, then

$$\sqrt{\left\{c^2e^2+(\lambda_1-be)^2\right\}k_1{}^2}=1, \tag{4.26}$$

and

$$k_1=\pm\frac{1}{\sqrt{c^2e^2+(\lambda_1-be)^2}}. \tag{4.27}$$

Similarly

$$k_2=\pm\frac{1}{\sqrt{c^2e^2+(\lambda_2-be)^2}}. \tag{4.28}$$

Considering the positive options for k_1 and k_2:

$$k_1=\frac{1}{\sqrt{c^2e^2+(\lambda_1-be)^2}}, k_2=-\frac{1}{\sqrt{c^2e^2+(\lambda_2-be)^2}}. \tag{4.29}$$

Substituting (4.29) for k_1 and k_2 in (4.22) and (4.23), two new vectors are derived:

$$\mathbf{u}_1=\frac{1}{\sqrt{c^2e^2+(\lambda_1-be)^2}}\begin{bmatrix}ce\\ \lambda_1-be\end{bmatrix}=\begin{bmatrix}\frac{ce}{\sqrt{c^2e^2+(\lambda_1-be)^2}}\\ \frac{\lambda_1-be}{\sqrt{c^2e^2+(\lambda_1-be)^2}}\end{bmatrix}, \tag{4.30}$$

$$\mathbf{u}_2=-\frac{1}{\sqrt{c^2e^2+(\lambda_2-be)^2}}\begin{bmatrix}ce\\ \lambda_2-be\end{bmatrix}=\begin{bmatrix}-\frac{ce}{\sqrt{c^2e^2+(\lambda_2-be)^2}}\\ -\frac{\lambda_2-be}{\sqrt{c^2e^2+(\lambda_2-be)^2}}\end{bmatrix}. \tag{4.31}$$

Matrix $\mathbf{U}$ is composed of the two vectors in (4.30) and (4.31):

$$\mathbf{U}=\begin{bmatrix}\mathbf{u}_1 & \mathbf{u}_2\end{bmatrix}=\begin{bmatrix}\frac{ce}{\sqrt{c^2e^2+(\lambda_1-be)^2}} & -\frac{ce}{\sqrt{c^2e^2+(\lambda_2-be)^2}}\\ \frac{\lambda_1-be}{\sqrt{c^2e^2+(\lambda_1-be)^2}} & -\frac{\lambda_2-be}{\sqrt{c^2e^2+(\lambda_2-be)^2}}\end{bmatrix}. \tag{4.32}$$

Simplifying (4.18) as follows:

$$\lambda_1=\frac{be+cf+\sqrt{(be+cf)^2-4bcef+4c^2e^2}}{2}=\frac{be+cf+\sqrt{(be-cf)^2+4c^2e^2}}{2}, \tag{4.33}$$

$$\lambda_2=\frac{be+cf-\sqrt{(be+cf)^2-4bcef+4c^2e^2}}{2}=\frac{be+cf-\sqrt{(be-cf)^2+4c^2e^2}}{2}, \tag{4.34}$$

we get:

$$\lambda_1-be=\frac{-be+cf+\sqrt{(be-cf)^2+4c^2e^2}}{2}=\frac{-(be-cf)+\sqrt{(be-cf)^2+4c^2e^2}}{2}>0, \tag{4.35}$$

$$\lambda_2 - be = \frac{-be + cf - \sqrt{(be-cf)^2 + 4c^2e^2}}{2} = \frac{-(be-cf) - \sqrt{(be-cf)^2 + 4c^2e^2}}{2} < 0, \tag{4.36}$$

Since $ce > 0,\ \lambda_1 - be > 0,\ \lambda_2 - be < 0,$

Rewriting (4.32) leads to the following Matrix:

$$\mathbf{U} = \begin{bmatrix} \frac{ce}{\sqrt{c^2e^2+(\lambda_1-be)^2}} & -\frac{ce}{\sqrt{c^2e^2+(\lambda_2-be)^2}} \\ \frac{\lambda_1-be}{\sqrt{c^2e^2+(\lambda_1-be)^2}} & -\frac{\lambda_2-be}{\sqrt{c^2e^2+(\lambda_2-be)^2}} \end{bmatrix} = \begin{bmatrix} \frac{1}{\sqrt{1+\frac{(\lambda_1-be)^2}{c^2e^2}}} & -\frac{1}{\sqrt{1+\frac{(\lambda_2-be)^2}{c^2e^2}}} \\ \frac{1}{\sqrt{\frac{c^2e^2}{(\lambda_1-be)^2}+1}} & \frac{1}{\sqrt{\frac{c^2e^2}{(\lambda_2-be)^2}+1}} \end{bmatrix}. \tag{4.37}$$

From (4.35)

$$\begin{aligned}(\lambda_1 - be)^2 &= \left(\frac{-(be-cf) + \sqrt{(be-cf)^2 + 4c^2e^2}}{2}\right)^2 \\ &= \left(\frac{-\omega + \sqrt{\omega^2 + 4c^2e^2}}{2}\right)^2 \\ &= \frac{2\omega^2 - 2\omega\sqrt{\omega^2 + 4c^2e^2} + 4c^2e^2}{4} \\ &= \frac{\omega^2 + 2c^2e^2 - \omega\sqrt{\omega^2 + 4c^2e^2}}{2}.\end{aligned} \tag{4.38}$$

Suppose $be - cf = \omega$, then (4.39)

$$\begin{aligned}(\lambda_2 - be)^2 &= \left(\frac{-(be-cf) - \sqrt{(be-cf)^2 + 4c^2e^2}}{2}\right)^2 \\ &= \left(\frac{-\omega - \sqrt{\omega^2 + 4c^2e^2}}{2}\right)^2 \\ &= \frac{2\omega^2 + 2\omega\sqrt{\omega^2 + 4c^2e^2} + 4c^2e^2}{4} \\ &= \frac{\omega^2 + 2c^2e^2 + \omega\sqrt{\omega^2 + 4c^2e^2}}{2} \\ &= \left(\frac{\omega^2 + 2c^2e^2 + \omega\sqrt{\omega^2 + 4c^2e^2}}{2}\right)\left(\frac{\omega^2 + 2c^2e^2 - \omega\sqrt{\omega^2 + 4c^2e^2}}{\omega^2 + 2c^2e^2 - \omega\sqrt{\omega^2 + 4c^2e^2}}\right)\end{aligned}$$

$$= \frac{\left(\omega^2+2c^2e^2\right)^2-\left(\omega\sqrt{\omega^2+4c^2e^2}\right)^2}{2\left(\omega^2+2c^2e^2-\omega\sqrt{\omega^2+4c^2e^2}\right)}$$

$$= \frac{\omega^4+4c^2e^2\omega^2+4c^4e^4-\left(\omega^4+4c^2e^2\omega^2\right)}{2\omega^2+4c^2e^2-2\omega\sqrt{\omega^2+4c^2e^2}}$$

$$= \frac{4c^4e^4}{2\omega^2+4c^2e^2-2\omega\sqrt{\omega^2+4c^2e^2}}$$

$$= \frac{2c^4e^4}{\omega^2+2c^2e^2-\omega\sqrt{\omega^2+4c^2e^2}}. \tag{4.40}$$

Substituting $(\lambda_2-be)^2$ with $\frac{2c^4e^4}{\omega^2+2c^2e^2-\omega\sqrt{\omega^2+4c^2e^2}}$ and inserting that in (4.37):

$$\mathbf{U}=\begin{bmatrix} \frac{1}{\sqrt{1+\frac{(\lambda_1-be)^2}{c^2e^2}}} & -\frac{1}{\sqrt{1+\frac{(\lambda_2-be)^2}{c^2e^2}}} \\ \frac{1}{\sqrt{\frac{c^2e^2}{(\lambda_1-be)^2}+1}} & \frac{1}{\sqrt{\frac{c^2e^2}{(\lambda_2-be)^2}+1}} \end{bmatrix}$$

$$=\begin{bmatrix} \frac{1}{\sqrt{1+\frac{\omega^2+2c^2e^2-\omega\sqrt{\omega^2+4c^2e^2}}{2}\cdot\frac{1}{c^2e^2}}} & -\frac{1}{\sqrt{1+\frac{2c^4e^4}{\omega^2+2c^2e^2-\omega\sqrt{\omega^2+4c^2e^2}}\cdot\frac{1}{c^2e^2}}} \\ \frac{1}{\sqrt{c^2e^2\cdot\frac{2}{\omega^2+2c^2e^2-\omega\sqrt{\omega^2+4c^2e^2}}+1}} & \frac{1}{\sqrt{c^2e^2\cdot\frac{\omega^2+2c^2e^2-\omega\sqrt{\omega^2+4c^2e^2}}{2c^4e^4}+1}} \end{bmatrix}$$

$$=\begin{bmatrix} \frac{1}{\sqrt{1+\frac{\omega^2+2c^2e^2-\omega\sqrt{\omega^2+4c^2e^2}}{2c^2e^2}}} & -\frac{1}{\sqrt{1+\frac{2c^2e^2}{\omega^2+2c^2e^2-\omega\sqrt{\omega^2+4c^2e^2}}}} \\ \frac{1}{\sqrt{1+\frac{2c^2e^2}{\omega^2+2c^2e^2-\omega\sqrt{\omega^2+4c^2e^2}}}} & \frac{1}{\sqrt{1+\frac{\omega^2+2c^2e^2-\omega\sqrt{\omega^2+4c^2e^2}}{2c^2e^2}}} \end{bmatrix}. \tag{4.41}$$

From (4.41) twisting angle θ can be calculated as follows:

$$\cos\theta=\frac{1}{\sqrt{1+\frac{\omega^2+2c^2e^2-\omega\sqrt{\omega^2+4c^2e^2}}{2c^2e^2}}},\ \sin\theta=\frac{1}{\sqrt{1+\frac{2c^2e^2}{\omega^2+2c^2e^2-\omega\sqrt{\omega^2+4c^2e^2}}}}. \tag{4.42}$$

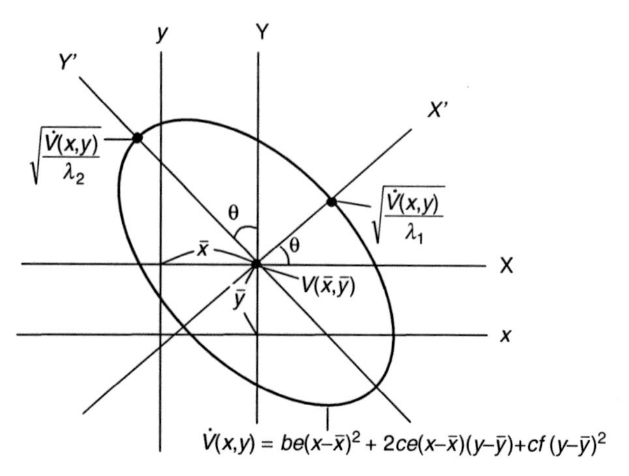

Fig. 4.5 General Image of an elliptical orbit

4.3.3 General Image of an Elliptical Orbit

Based on the foregoing mathematical analysis, an orbit of competing technologies x, y with certain interplay conditions can be depicted as follows, and general image of an elliptical orbit under certain $\dot{V}$ condition (in case of $bf > ce$) can be illustrated as Fig. 4.5:

Interplay conditions

$$\dot{x} = x(a - bx - cy)$$
$$\dot{y} = y(d - ex - fy).$$

Orbit

$$\dot{V}(x,y) = be(x-\bar{x})^2 + 2ce(x-\bar{x})(y-\bar{y}) + cf(y-\bar{y})^2,$$

$$\frac{X'^2}{\left(\sqrt{\frac{\dot{V}(x,y)}{\lambda_1}}\right)^2} + \frac{Y'^2}{\left(\sqrt{\frac{\dot{V}(x,y)}{\lambda_2}}\right)^2} = 1,$$

$$\lambda_1 = \frac{be + cf + \sqrt{(be+cf)^2 - 4bcef + 4c^2e^2}}{2},$$

$$\lambda_2 = \frac{be + cf - \sqrt{(be+cf)^2 - 4bcef + 4c^2e^2}}{2}.$$

Twisting angle

$$\cos\,\theta = \frac{1}{\sqrt{1 + \frac{\omega^2 + 2c^2e^2 - \omega\sqrt{\omega^2 + 4c^2e^2}}{2c^2e^2}}},$$

$$\sin\,\theta = \frac{1}{\sqrt{1+\frac{2c^2e^2}{\omega^2+2c^2e^2-\omega\sqrt{\omega^2+4c^2e^2}}}},$$

where $\omega = be - cf$.

Given the technologies x and y compete each other constructing a complex orbit following the foregoing conditions, Fig. 4.5 provides suggestive idea for optimal policy option.

4.4 Orbit for Substitution: Policy Option in a Complex Orbit

While there exists variety of orbits in predator–prey systems [16, 17] given such transition as anticipated in shift from analog to digital TV broadcasting analyzed in Sect. 4.2, following boundary conditions are imposed.

Provided that a species x is a preceding species and, later on, a species y appears and steadily succeeds x, and finally substitutes for x, an orbit for y substitute for x can be developed as follows:

In general interaction of two competing species x and y, interaction coefficients in (4.1) $\alpha_{xy}\left(=\frac{cd}{af}\right)$ and $\alpha_{yx}\left(=\frac{ae}{bd}\right)$ are

$$\frac{1 \geqslant \alpha_{xy} > .0, \text{ i.e. } a/c \geqslant d/f}{1 \geqslant \alpha_{yx} > .0, \text{ i.e. } d/e \geqslant a/b}. \tag{4.43}$$

A condition of the initial stage just before the substantial emergence of species y can be depicted as $\frac{\partial \dot{V}}{\partial y} = 0$

i.e.

$$\begin{aligned} \frac{\partial \dot{V}}{\partial y} &= 2ce(x-\bar{x}) + 2cf(y-\bar{y}) = 0, \\ e(x-\bar{x}) &+ f(y-\bar{y}) = 0, \\ ex + fy &= e\bar{x} + f\bar{y} = d. \end{aligned} \tag{4.44}$$

Equation (4.1) suggests that this is equivalent to $\frac{\dot{y}}{y} = 0$.

Similarly, a condition of the stage when y totally substitutes for x can be depicted as $\frac{\partial \dot{V}}{\partial x} = 0$

i.e.

$$bx + cy = b\bar{x} + c\bar{y} = a. \tag{4.45}$$

This is equivalent to $\frac{\dot{x}}{x} = 0$.

The isoclines of the above dynamism that satisfy conditions enumerated by (4.43)–(4.45) can be illustrated as Fig. 4.6.

Under the condition that $x, y > 0$ (hence $\bar{x}, \bar{y} > 0$) and given the case when y succeeds x, $d/e \geqslant a/b$ as well as $a/c \geqslant d/f$ and hence $b/e \geqslant a/d \geqslant c/f$ conditions should be satisfied which lead to $bf > ce$.

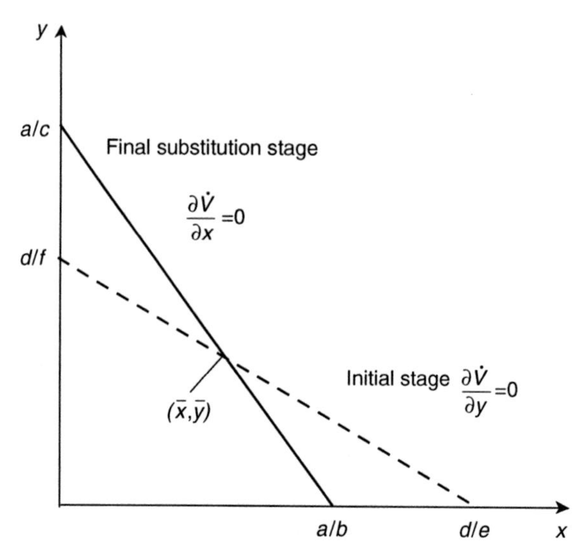

Fig. 4.6 Isoclines for two dimensional Lotka–Volterra equations

These conditions imply the followings with respect to an orbit of two-dimensional Lotka–Volterra equations for y substitutes for x:

(a) The orbit follows an elliptical orbit,
(b) Relationship between x and y is the case of stable coexistence, and
(c) Equilibrium point of this coexistence $V(x,y) = V(\bar{x},\bar{y})$.

Under the above dynamism, given the situation when y totally substitutes for x at the final substitution stage, $\bar{x} = 0$ and hence $a/c = d/f$ should be satisfied which implies the followings with respect to an orbit in the period starting from the initial stage when y first invades into $x(\partial\dot{V}/\partial y = 0)$ and ending final substitution stage when y totally substitutes for x $(\partial\dot{V}/\partial x = 0)$:

(a) The orbit $V(x,y)$ moves from $V(d/e, 0)$ to $V(0, a/c)$, and
(b) The interaction coefficient $\alpha_{xy}(=cd/df)=1$ (see (4.1)) while $\alpha_{yx}(=ae/bd)<1$.

(This implies y's invasion power into x territory is stronger than that of x into y.)

Thus, isoclines for two-dimensional Lotka–Volterra equations under substitution orbit can be illustrated as Fig. 4.7.

As general two-dimensional game suggests, after certain game, $\dot{V}$ stagnates steadily and reaching $\dot{V} = 0$,[4] by synchronizing Figs. 4.5 and 4.7, general image of an elliptical orbit for substitution can be illustrated as Fig. 4.8.

Under the condition when $V(x, y)$ shifts to the state of equilibrium with respect to y substitution for x with certain constant pace,[5] an orbit of Fig. 4.5 can be projected to respective time trend of x and y as illustrated in Fig. 4.9. In this case, time scale

[4] This implies the state of equilibrium with respect to y substitution for x, and does not imply the termination of y or x increase.

[5] By differentiating $\dot{V}(x, y)$ in (4.7) by time t we obtain

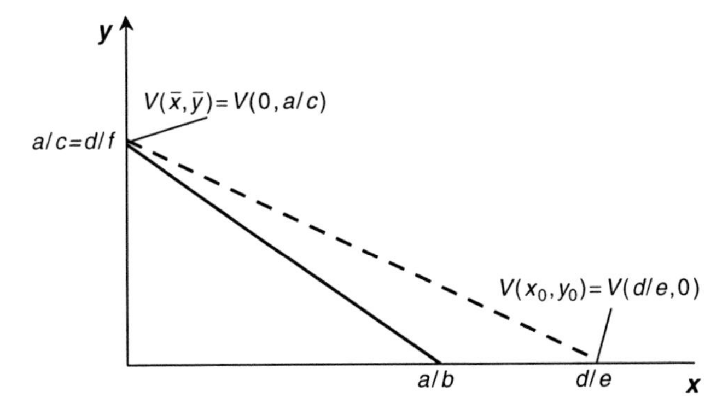

Fig. 4.7 Isoclines for two-dimensional Lotka–Volterra equations under substitution orbit

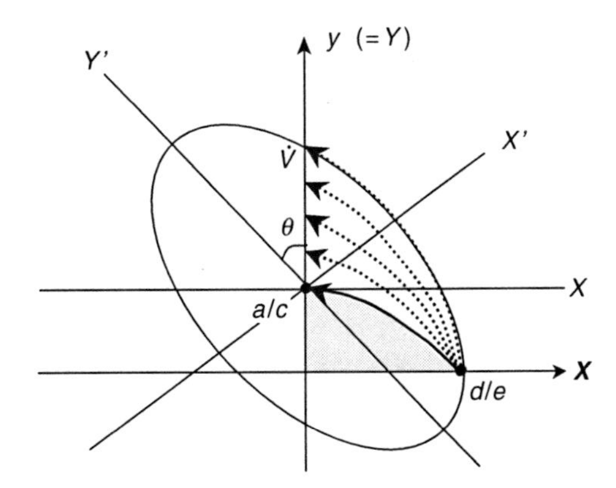

Fig. 4.8 General image of an elliptical orbit for substitution

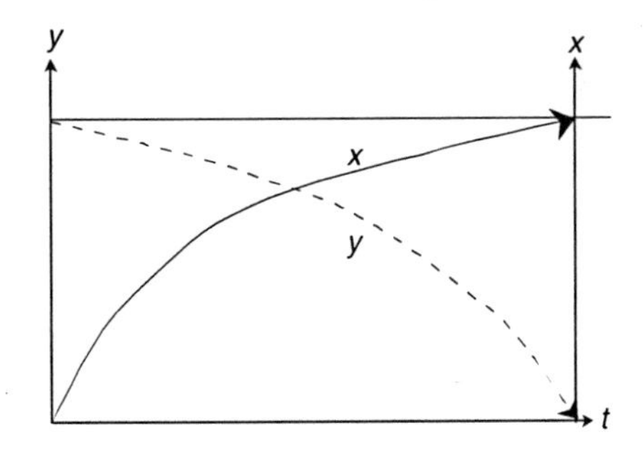

Fig. 4.9 Trend in x and y under y substitutes for x condition
[a] x and y indicate the diffusion ratio

$$\frac{\mathrm{d}\dot{V}(x,y)}{\mathrm{d}t} = -2\left\{ex(a-bx-cy)^2 + cy(d-ex-fy)^2\right\} \leqslant 0$$

$$\frac{\mathrm{d}\dot{V}(x,y)}{\mathrm{d}t} = 0 \text{ when } \dot{V}(x,y) = \dot{V}(\bar{x},\bar{y}).$$

This suggests that an orbit $\dot{V}(x,y)$ shifts toward the equilibrium point $\dot{V}(\bar{x},\bar{y})$ with a pace of $g(\equiv -2\left\{ex(a-bx-cy)^2 + cy(d-ex-fy)^2\right\})$. Given a constant g, $\dot{V}(x,y)$ can be depicted as $\dot{V}(x,y) = \dot{V}_0(x,y)e^{gt}$ where $\dot{V}_0(x,y)$ indicates initial change.

t differs whether it is "market time scale" or "innovation time scale." In case of the latter scale, diffusion of new innovation y is generally slow, but if it is less complex, encompasses a possibility of rapid diffusion which cannot be expected under "market time scale" as is observed in the transition from monochrome TV to color TV (Fig. 4.1). Provided that diffusion y is expected to be forwarded promptly as reviewed in Sect. 4.2.1 (see Fig. 4.4), y should be concave in an innovative time scale and this could be expected given y is less complex by making every effort to be human friendly one [21].

Figure 4.10 compares orbits between exponential function, logistic (epidemic) function and Lotka–Volterra function (see Sect. 4.2.2).[6] Under the foregoing condition that y is less complex and extremely human friendly one, Lotka–Volterra orbit could display concave and higher growth rate at the initial stage. Given the less complex innovation, this can be possible because of the maturity of growth condition, and due not only to the substitution proceeds for existing competitive species (x) under pure competition without any institutional constraints, but also to customer realizes potential benefit of y. This is quite similar to the diffusion orbit of transition from the cellular telephones to mobile internet access service as reviewed in Sect. 4.2.1 (see Fig. 4.3).

As analyzed in Fig. 4.10, contrary to the logistic growth orbit, Lotka–Volterra orbit encompasses a possible orbit with higher dependency on new technology which substitutes for old one from the early stage of its introduction. As reviewed in Sects. 4.1 and 4.3, given the slightly more "successful species" existence, Lotka–Volterra orbit represents substitution and diffusion orbit under the conditions of pure

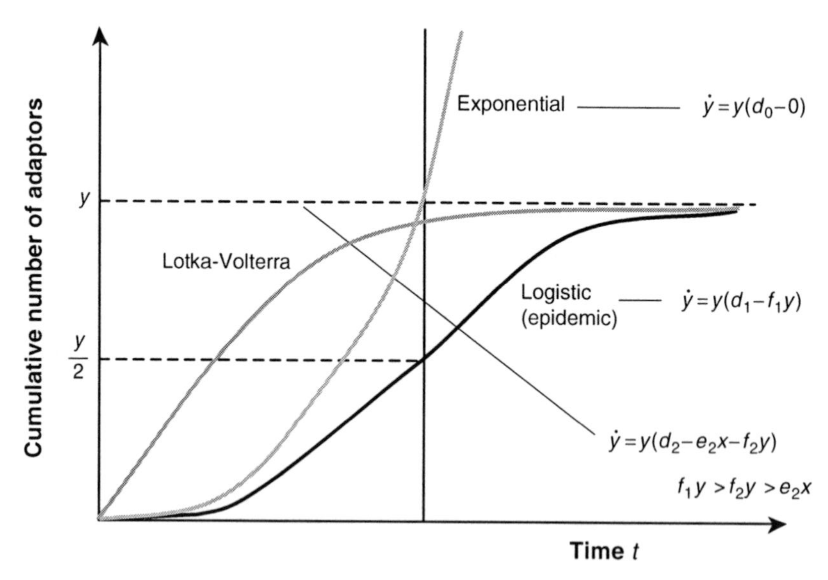

Fig. 4.10 Comparison between exponential function, logistic function and Lotka–Volterra function

[6] See [6, 14–16] for comparison of diffusion orbit.

competition between competing species just by functions of respective species with fair information, without fear of substitution, a reluctance to pay the cost of switching, and barriers within the manufacturing industry.

Therefore, for the acceleration of the diffusion process, factors separating the two orbits, logistic growth orbit and Lotka–Volterra orbit, should be removed. In order to lead Lotka–Volterra orbit to such expected orbit, every efforts to enhance the following three factors governing the orbit y such as, d (maximum diffusion scale), k_y (carrying capacity), and α_{yx} (interaction coefficient) in (4.1) should be focused:

4.4.1 Maximum Diffusion Scale: d

d indicates the maximum diffusion of innovative goods in the market. Since the actual diffusion of innovative goods greatly depends on the elasticity of the potential users, it is crucial to take measures to stimulate the elastic reaction of the potential users to the innovative goods in order to accelerate the diffusion process.

From users' point of view, major factors that prevent them from adopting or switching to new technology are a lack of information about the innovation, that is, they do not know what it is and how to use it, as well as a reluctance to pay the cost of switching. In this sense, efforts by supply side of the innovation such as intensive advertising for publicizing the merits of the innovation, providing user friendly interface, and public education are considered as very effective to push reluctant users to adopt the innovation spontaneously.

4.4.2 Carrying Capacity: $\mathbf{k_y}$

k_y indicates the ultimate upper limit (carrying capacity) of the adoptions of innovative goods. As Watanabe and Kondo et al. [18] verified by analyzing the diffusion process of cellular telephones in Japan, since the diffusion process of the innovation which creates new functions during the course of its interaction with users can be well modeled by logistic growth within a dynamic carrying capacity, the potential of k_y can be regarded as much depending on the potential of the innovation towards multi-functionality.

There are mainly two factors to achieve multi-functionality: one is the R&D efforts by suppliers of the innovation and the other is the network externality of the innovation together with the elasticity of the market. It is rather obvious that new innovation obtained from the R&D efforts by suppliers have the possibility to add additional or completely new functionality to the original innovation, leading to achieving multi-functionality.

The other factor depends on the nature of the innovation. If the innovation has a feature of network externality, that is, if the innovation possesses a feature that the more it is deployed, the greater its value to the adopters, it has the potential to attract more and more users as well as other industries, assuming that the market is elastic enough to adopt the innovation. Broader interaction with users and other industries stimulate adding new functionalities to the original innovation.

4.4.3 Interaction Coefficient: α_{yx}

α_{yx} indicates the intensity of competition between the existing goods (x) in the market and newly entering innovative goods (y). The value differs depending on the relationship between the two innovative goods: complementary, substitute, or fully competitive.

When the relationship is *complementary*, y's diffusion proceeds rather quickly if x has already established some market share. Since y enhances the value of x, the diffusion of x and y is expected to grow mutually. When y substitutes x, it means that y's invasion power into x territory is stronger than that of x into y.

4.5 Conclusion

In light of the increasing significance of timely introduction of emerging new technologies that substitute for existing technology for enhancing a nation's international competitiveness in a globalizing economy, this chapter, focusing on Japan's transition from manufacturing technology to IT, analyzes substitution orbits of two competitive innovations.

Prompted by a sophisticated balance of the co-evolution process in an ecosystem, particularly of the complex interplay between competition and cooperation, an application of Lotka–Volterra equations that analyze these sophisticated balance in an ecosystem, is conducted for reviewing the substitution orbits of Japan's monochrome to color TV system, fixed telephones to cellular telephones, and cellular telephones to mobile internet access service, and analog to digital TV broadcasting.

On the basis of the comparative assessment by using synthesized two dimensional Lotka–Volterra equations for substitution and general logistic growth equation, following findings are obtained:

(1) Lotka–Volterra equations for substitution are useful for identifying an optimal orbit of competitive innovations with complex orbit.
(2) It is particularly useful for assessing policy options from the view point of effectiveness in controlling parameters for leading to expected orbit.
(3) Under the global paradigm shift from an industrial society to an information society, given the government target to accomplish a rapid shift from traditional technology to new technology, particularly IT driven new technology within a limited period, shifting scenario should be accelerated in line with Lotka–Volterra orbit with optimal coefficients.

(4) In order to accomplish this orbit, every efforts should be accelerated in removing factors separating the two orbits between logistic growth orbit and Lotka–Volterra orbit including:

    (a) A lack of information about new technologies expected to be substituted for traditional one,
    (b) Fear of substitution and a reluctances to pay the cost of switching from traditional to new technology, and
    (c) Barriers to prompt shift to new technology within producers, distributors and customers.

(5) In addition, in order to accelerate such shift with a higher pace of Lotka–Volterra orbit, with the understanding that IT's specific functionality is formed through dynamic interaction with institutional systems [19], efforts should be focused on maximizing IT's self-propagation behavior.

Considering that while hasty shift sometimes accomplishes nothing, delayed shift can result in a loss of national competitiveness, identification of the optimal diffusion and substitution orbit is essential, thus, the foregoing approach is very useful in identifying policy options for the diffusion orbit of competitive innovations with complex interplay.

Further mathematical attempts aiming at broader application of possible orbit for substitution, together with empirical analyses taking "success stories" in smooth substitution and rapid diffusion such as mobile internet access service on factors, conditions and systems enabled them rapid substitution and diffusion are expected to be undertaken.

References

1. A. Gruebler, Technology and Global Change (Cambridge University Press, Cambridge, 1998)
2. M. Bauer (ed.), Resistance to New Technology (Cambridge University Press, Cambridge, 1995)
3. C.M. Christensen, The Innovator's Dilemma (Harvard Business School Press, Cambridge, 1997)
4. J.F. Moore, Predators and Prey: A New Ecology of Competition (Harvard Business Review, May–June 1993) 75–83
5. H. Binswanger and V. Ruttan, Induced innovation: technology, institutions, and development (John Hopkins University Press, Baltimore, 1978)
6. V.W. Ruttan, Technology, Growth, and Development – An Induced Innovation Perspective (Oxford University Press, New York, 2001)
7. R.R. Nelson and B.N. Sampat, Making sense of institutions as a factor shaping economic performance, Journal of Economic Behavior & Organization 44, No. 1 (2001) 31–54
8. M. Sonis, Dynamic choice of alternatives, innovation diffusion, and ecological dynamics of the Volterra–Lotka model, London Papers in Regional Science 14 (1984) 29–43
9. J. Hofbander and K. Sigmund, The Theory of Evolution and Dynamical Systems (Cambridge University Press, Cambridge, 1988)
10. E.H. Kerner, On the Volterra–Lotka Principle, Bulletin of Mathematical Biophysics 23 (1961) 133–149

11. A.M. Noll, The evolution of television technology, in D. Gerbang (ed.), The Economics, Technology and Content of Digital TV (Kluwer, Norwell, MA, 1999) 3–17
12. R. Parker, The economics of digital TV's future, in D. Gerbang (ed.), The Economies, Technology and Content of Digital TV (Kluwer, Norwell, MA, 1999) 197–213
13. Ministry of Public Management, Home Affairs, Posts and Telecommunications (MPMPT), Report on the Next Generation Broadcasting Technologies (MPMPT, Tokyo, 2001)
14. J.A. Hart, Digital television in Europe and Japan, in D. Gerbang (ed.), The Economics, Technology and Content of Digital TV (Kluwer, Norwell, MA, 1999) 287–314
15. Advisory Committee on Digital Terrestrial Broadcasting (ACDTB), Construction of New Terrestrial Broadcasting System (ACDTB, Tokyo, 1998)
16. F. Scudo and J. Ziegler, The golden age of theoretical ecology, 1923–1940, in Lecture Notes in Biomathematics 22 (Springer, Berlin, 1978)
17. A.A. Harms and E.M. Krenciglowa, An extended Lotka–Volterra model for population – energy systems, Paper presented at the International Symposium on Energy and Ecological Modelling (Louisville, Kentucky, 1981)
18. C. Watanabe, R. Kondo, N. Ouchi and H. Wei, Formation of IT features through interaction with institutional systems, Technovation 23, No. 3 (2003) 205–219
19. C. Antonelli, New information technology and the evolution of the industrial organization of the production of knowledge, in D. Lamberton et al. (ed.) Information and Organization (Elsevier Science, New York, 1999) 263–284
20. Ministry of Public Management, Home Affairs, Posts and Telecommunications (MPMPT), White Paper 2001 on Communications in Japan (MPMPT, Tokyo, 2001)
21. P.A. Geroski, Models of technology diffusion, Research Policy 29, Nos. 4–5 (2000) 603–625

Chapter 5
Impacts of Functionality Development on Dynamism Between Learning and Diffusion of Technology

Abstract Under a long lasting economic stagnation, since significant increase in R&D investment has become difficult, practical solution could be found in systems approach maximizing the effects of innovation as a system by making full utilization of potential resources of innovation. At the same time, under the increasing significance of information technology (IT) in an information society which emerged in the 1990s, functionality development has become crucial for stimulating a self-propagating nature of IT driven innovation.

Stimulated by these understandings and prompted by a concept of institutional innovation, this chapter attempts to analyze the interacting dynamism of innovation in a comprehensive and organic system. Theoretical analysis and empirical demonstration are attempted focusing on dynamism between learning and diffusion of technology taking Japan's PV development, which follows the similar trajectory of IT's functionality development, over the last quarter century.

The effects of functionality decrease on learning coefficient and consequent impacts on technology diffusion and its dynamic carrying capacity are analyzed. Fear of a vicious cycle between functionality decrease, deterioration of learning, stagnation of technology diffusion and its carrying capacity in long run is demonstrated. Thereby, the significance of institutional dynamism leading to a dynamic interaction between learning, diffusion and spillover of technology is identified.

Reprinted from *Technovation* 24, No. 8, C. Watanabe and B. Asgari, Impacts of Functionality Development on the Dynamism Between Learning and Diffusion of Technology, pages: 651–664, copyright (2004), with permission from Elsevier.

5.1 Introduction

While technological innovation plays a significant contribution to socio-economic development, under a long lasting economic stagnation, the stagnation of technology development has become a crucial structural problem common to all advanced countries [1]. Similarly, Japan has been suffering from a collapse of its long lasting

C. Watanabe, *Managing Innovation in Japan: The Role Institutions Play in Helping or Hindering how Companies Develop Technology*, DOI: 10.1007/978-3-540-89272-4_5,

virtuous cycle between technology development and economic growth [2] leading to a vicious cycle between economic stagnation and the stagnation of R&D investment. Under such circumstances, significant increase in R&D investment has become difficult requirements which increases the significance of the systems approach maximizing the effects of innovation as a system [3]. At the same time, under the increasing significance of information technology (IT) in an information society which emerged in the 1990s, functionality development has become crucial for stimulating a self-propagating nature of IT driven innovation [4].

Prompted by these understandings, this chapter attempts to analyze the interacting dynamism of innovation in a comprehensive and organic system. As postulated by Ruttan [5] innovation should be recognized as a very subtle entity subject to conditions of institutional systems. Therefore, theoretical analysis and empirical demonstration are attempted focusing on a dynamism between learning and diffusion of technology taking Japan's PV development over the last quarter century. PV development trajectory is taken as it follows the similar trajectory of IT's functionality development [6].

In line with the foregoing economic as well as technology stagnation, it is generally anticipated that functionality development in Japan's high-technology industry has decreased. The new functionality development created by electrical machinery, which shares one third of Japan's whole R&D expenditure, was dramatically exhausted in the 1990s [7].

Such a decrease in the functionality development inevitably results in a decrease in learning effects [8]. As Cohen and Levinthal [9] postulated, learning is cumulative and cumulative learning stimulates assimilation of spillover knowledge which inevitably induces distribution of technology. Furthermore, as Watanabe et al. [7] demonstrated, functionality development concept can be materialized by correlating technology elasticity to sales as well as logistic growth within a dynamic carrying capacity which depicts diffusion of technology and its dynamic carrying capacity that represents state of the functionality development [7]. Thus, decrease in functionality development is anticipated to lead to a vicious cycle between decrease in learning effects (which can be measured by decrease in learning coefficient), stagnation of technology diffusion and its dynamic carrying capacity (which represents further decrease in functionality).

Following Arrow's pioneer postulate on "learning-by-doing" [10], while a number of works analyzed the mechanism of learning and its effects [9, 11], none has analyzed dynamic hysteresis of learning coefficient. Similarly, since Rogers' pioneer work on diffusion of innovations [12], a number of works analyzed diffusion process of technology [13, 14], as well as governing factors of the diffusion trajectory [15, 16], none has linked dynamic behavior of learning coefficient with trajectory of technology diffusion and its dynamic carrying capacity in a virtuous or vicious cycle perspective.

In light of the foregoings, this chapter, by means of theoretical analysis and empirical demonstration, attempts to analyze the effects of functionality decrease on learning coefficient and consequent impacts on technology diffusion and dynamic carrying capacity are analyzed. Fear of a vicious cycle between functionality decrease, deterioration of learning, stagnation of technology diffusion and carrying capacity in long run is demonstrated.

Section 5.2 attempts to analyze the dynamic behavior of learning coefficient by constructing a mathematical model and empirical demonstration. Section 5.3 links learning and diffusion of technology by developing this mathematical model. Section 5.4 provides an interpretation of these analyses by elaborating an institutional dynamism leading to a dynamic construction between learning, diffusion and spillovers of technology. Section 5.5 briefly summarizes the findings obtained from the analyses and extracts policy implications for effective utilization of potential resources of innovation.

5.2 Dynamic Behavior of Learning Coefficient

Learning exercise is a result of cumulative efforts and it is a long-range strategic concept rather than a short-term tactical concept. It represents the combined effects of a large number of factors and cumulative efforts.

Operating in competitive markets makes individuals, firms, industries and nations do better. This motivation is at the heart of the learning exercise phenomenon and subsequent learning effects. Price is the most important measure of performance for this motivation [17] and returns of consequent cumulative efforts, generally expressed by cumulative production.

Thus, learning effects can be captured by the following equation:

$$P = B \cdot Y^{*-\lambda} \tag{5.1}$$

where P, prices, B, scale factor; $Y^* = \Sigma Y$, cumulative production (Y, production)[1]; and λ (> 0), learning coefficient.

Taking logarithm of (5.1)

$$\ln P = \ln B - \lambda \ln Y^*. \tag{5.2}$$

Differentiating both sides of (5.2) with respect to time t,

$$-\lambda = \frac{\frac{d}{dt} \ln P}{\frac{d}{dt} \ln Y^*}. \tag{5.3}$$

In case of innovative goods, prices can be depicted by a function of time t as demonstrated by the decreasing trend in PV prices in Japan which is clearly illustrated in Fig. 5.1.

[1] Given the production at time t, Y_t, cumulative production at time t, Y_t^*, can be measured as follows:

$$Y_t^* = Y_{t-l_t} + (1-\rho)Y_{t-1}^*$$

where l_t, lead time between production and operation; and ρ, depreciation rate.

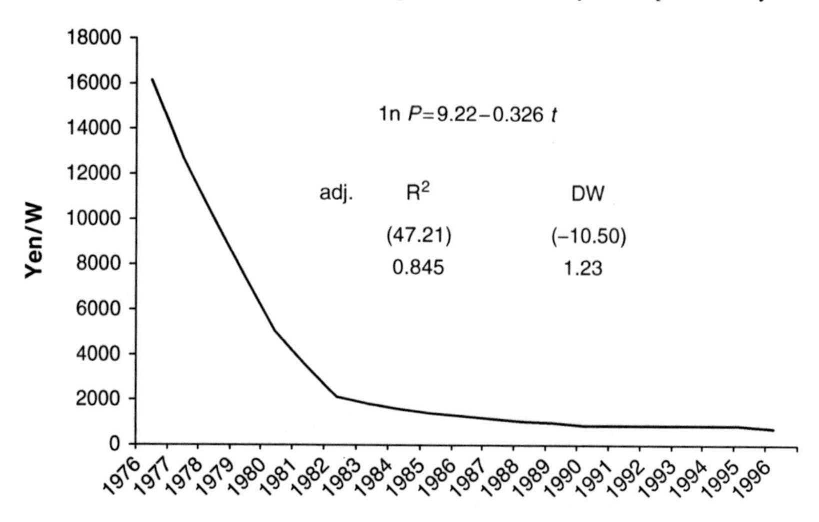

Fig. 5.1 Trend in PV prices in Japan (1976–1996) – Yen/W at 1985 fixed prices

$$P = B' \mathrm{e}^{-\eta t}, \tag{5.4}$$

where B', scale factor; η, coefficient; and t, time trend.

Figure 5.1 demonstrates statistically significant correlation between time t and prices P.

From (5.4) coefficient η can be obtained by the following equation:

$$\frac{\mathrm{d}\ln P}{\mathrm{d}t} = -\eta. \tag{5.5}$$

Trajectory of diffusion process of Y^* can be depicted by the following epidemic function:

$$\frac{\mathrm{d}Y^*}{\mathrm{d}t} = bY^* \left(1 - \frac{Y^*}{K}\right), \tag{5.6}$$

where b, coefficient; and K, carrying capacity.

Equation (5.6) can be developed to

$$\frac{\mathrm{d}\ln Y^*}{\mathrm{d}t} = b \left(1 - \frac{Y^*}{K}\right). \tag{5.7}$$

Provided that diffusion process of Y^* follows a trajectory depicted by a logistic growth function within a dynamic carrying capacity (LFDCC), Y^* can be depicted as follows (see mathematical development in Appendix):

$$Y^* = \frac{K_k}{1 + a\mathrm{e}^{-bt} + \frac{a_k}{1-b_k/b}\mathrm{e}^{-b_k t}}, \tag{5.8}$$

where $K = \frac{K_k}{1+a_k \mathrm{e}^{-b_k t}}$; K_k, ultimate carrying capacity; and a_k and b_k, coefficients.

Substituting Y^* in the right hand side of (5.7) for a trajectory depicted by (5.8):

$$\frac{\mathrm{d}\ln Y^*}{\mathrm{d}t} \approx b\left(1 - \frac{K_k/K}{1+a\mathrm{e}^{-bt} + \frac{a_k}{1-b_k/b}\mathrm{e}^{-b_k t}}\right)$$
$$= b'\left(1 - \frac{\varphi}{1+a\mathrm{e}^{-bt} + \frac{a_k}{1-b_k/b}\mathrm{e}^{-b_k t}}\right), \tag{5.9}$$

where b', adjusted coefficient, and ϕ (< 1), adjustment coefficient.

Substituting the right hand side of (5.5) and (5.9) for relevant factors of (5.3) under certain conditions when $\varphi \cdot \left(a\mathrm{e}^{-bt} + \frac{a_k}{1-b_k/b}\mathrm{e}^{-b_k t}\right) \ll$ learning coefficient λ can be approximated by the following equation:

$$\lambda = \frac{\eta}{b'}\left(\frac{1}{1 - \frac{\varphi}{1+a\mathrm{e}^{-bt} + \frac{a_k}{1-b_k/b}\mathrm{e}^{-b_k t}}}\right)$$
$$\approx \frac{\eta}{b'}\left\{1+\varphi\left(1-\left(a\mathrm{e}^{-bt} + \frac{a_k}{1-b_k/b}\mathrm{e}^{-b_k t}\right)\right)\right\}$$
$$= \frac{\eta}{b'}\left\{(1+\varphi) - \varphi\left(a\mathrm{e}^{-bt} + \frac{a_k}{1-b_k/b}\mathrm{e}^{-b_k t}\right)\right\} \tag{5.10}$$
$$\equiv \varphi_1 - \varphi_2\left(a\mathrm{e}^{-bt} + \frac{a_k}{1-b_k/b}\mathrm{e}^{-b_k t}\right),$$

where $\phi_1 = \frac{\eta}{b'}(1+\phi)$ and $\phi_2 = \frac{\eta}{b'}\phi$, coefficients (> 0).

Therefore, learning coefficient λ can be depicted by the following general equation:

$$\lambda = \alpha - \beta\mathrm{e}^{-\gamma t}, \tag{5.11}$$

where α, β, and γ are positive coefficients.

Equation (5.10) suggests that a coefficient γ is a function depicted by the following function:

$$\gamma = \gamma\left((a,b), \left(\frac{a_k}{1-b_k/b}, b_k\right)\right). \tag{5.12}$$

The second term in (5.12) is a function of factors governing dynamic carrying capacity and reflecting functionality of the innovative goods examined [7]. Since this functionality decreases in long run [8], γ can be expressed by the following function:

$$\gamma = l - mt, \tag{5.13}$$

where coefficients l and m are positive values.

Therefore, λ can be expressed by the following equation:

$$\lambda = \alpha - \beta\mathrm{e}^{-(l-mt)t}. \tag{5.14}$$

Equation (5.14) indicates convex with its peak at time $t = \frac{l}{2m}$ when $\frac{d\lambda}{dt} = 0$. Thus, a trajectory of λ starts from the initial level $\alpha - \beta$ (when $t = 0$), continues to increase its level by the period $t = \frac{l}{2m}$ with its peak level $\lambda_{\max} = \alpha - \beta\, e^{-\frac{l^2}{4m}}$, and then changes to decreasing trend. At time $t = \frac{l}{m}$, its level decreases to the same level of initial period $(\alpha - \beta)$, and continues to decrease to the lower level than initial period as demonstrated in Fig. 5.2.

In order to demonstrate the significance of the foregoing general equation of learning coefficient with respect to broader applicability, and given this significance is demonstrated, also to identify the behavior of this coefficient, an empirical analysis is conducted by taking Japan's PV development trajectory over the period 1976–1999. The results are summarized in Table 5.1 and Figs. 5.3 and 5.4.

Table 5.1 summarizes comparison of learning coefficients without considering functionality decrease and by considering functionality decrease for Japan's PV development trajectory over the period 1976–1999.

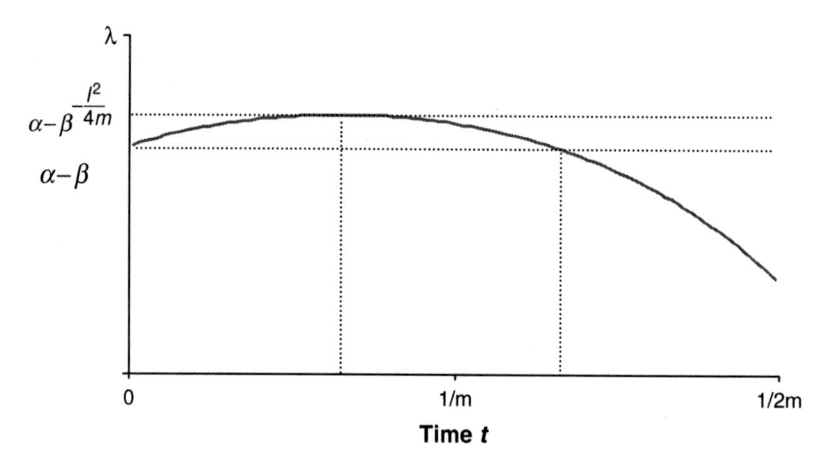

Fig. 5.2 Trajectory of learning coefficient

Table 5.1 Comparison of learning coefficient functions for Japan's PV development (1976–1999)

Without considering functionality decrease

$\ln P = 8.3000 - (0.3565 - 0.0088e^{-0.0089t}) \ln Y^*$

adj. R^2 DW

(239.06) (45.91) (4.04) (2.45) 0.990 0.81

$\lambda = 0.3565 - 0.0088e^{-8.9120t}$

Considering functionality decrease

$\ln P = 8.2927 - (0.3553 - 0.0086e^{-(0.0072-0.00011t)t}) \ln Y^*$

adj. R^2 DW

(290.86) (50.36) (4.17) (3.22) (5.79) 0.993 1.31

$\lambda = 0.3553 - 0.0086e^{-(0.0072-0.00011t)t}$

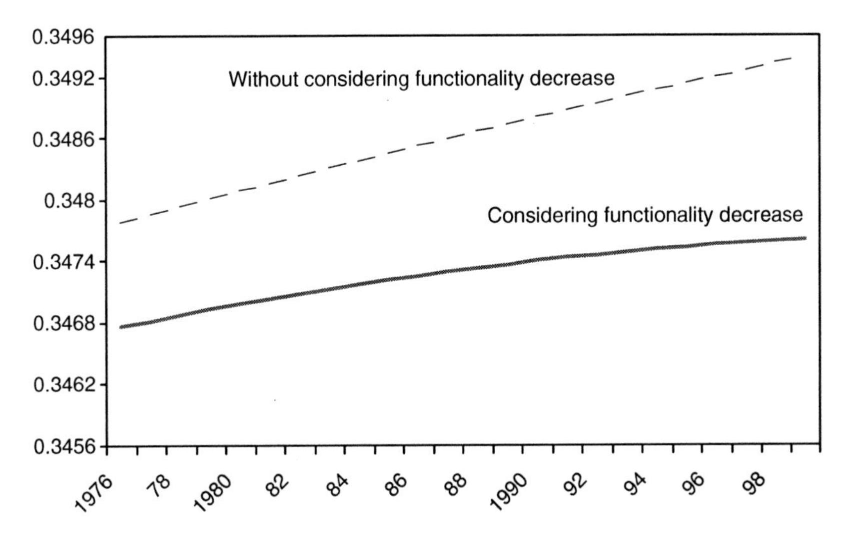

Fig. 5.3 Trends in learning coefficients in Japan's PV development (1976–1999)

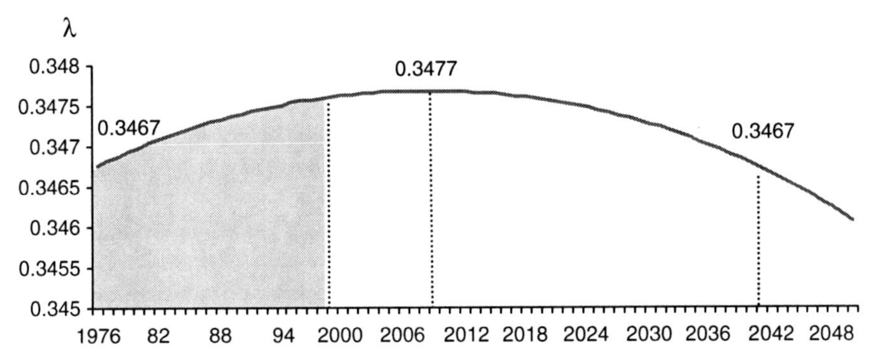

Fig. 5.4 Estimate of the future trajectory of learning coefficient in Japan's PV development (1976–2050)

Looking at Table 5.1 we note that learning coefficient function considering functionality decrease indicates statistically more significant than that of without considering functionality decrease. This result demonstrates that the effects of functionality decrease can not be overlooked in long run.

On the basis of the foregoing analysis, the learning coefficient for Japanese PV development is estimated as follows:

$$\lambda = 0.3553 - 0.0086\,\mathrm{e}^{-(0.0072-.00011t)t}. \tag{5.15}$$

Figure 5.3 demonstrates the trends in learning coefficients measured by this learning coefficient function by comparing with a trend estimated by a function without considering functionality decrease.

Table 5.2 Learning coefficients of PV development in leading Japanese PV firms

(1980–1990): Aggregate average of eight leading firms
Model: $P = A \cdot Y^{*^{-\lambda}}$ $\lambda = 0.347$
$\ln P = 3.609 - 0.347 \ln Y^{*}$ adj. R^2 DW
(73.40) (−22.80) 0.981 1.42
where P, solar cell production price (fixed price); and Y^{*}, cumulative solar cell production
The eight firms are: Sanyo Electric Co., Ltd., Kyocera Corp., Sharp Corp., Kaneka Corp., Fuji Electric Co., Ltd., and Hitachi, Ltd.

Source: Watanabe et al. (2001) [6]

Looking at Fig. 5.3 we note that the learning coefficient of Japan's PV development over the last quarter century indicates approximately 0.347 with sustaining slight increasing trend which corresponds to the learning coefficient of Japan's leading PV firms as demonstrated in Table 5.2.

On the basis of these analyses and evaluations, learning coefficient function based on the general equation driven by an approximation of LFDCC and considering functionality decrease can be considered well reflecting the trend in learning coefficient of the development trajectory of innovative goods. Figure 5.3 indicates that the trend measured without considering functionality decrease tends to demonstrate higher coefficient value than that of measured by considering functionality decrease.

Based on the foregoing assessment with respect to broad applicability of the learning coefficient function, Fig. 5.4 estimates the future trajectory of learning development in long run until 2050.

Looking at Fig. 5.4 we note that the learning coefficient demonstrates a convex trend with its peak at 0.3477 in 2009 (at time $t = \frac{l}{2m}$ when $\frac{d\lambda}{dt} = 0$). Since then the trend changes to decreasing trend. In 2041 its level reaches a level equal to its initial level (0.3467) and continues to decrease to the lower level than initial level. This clearly demonstrates that the significance of trajectory of learning coefficient considering functionality decrease in long run. The trend estimated in Fig. 5.3 demonstrates a partial section of the trend in the learning coefficient before reaching its peak.

5.3 Learning and Diffusion of Technology

The analysis in Sect. 5.2 demonstrates the broad applicability of the learning coefficient function driven by LFDCC and considering functionality decrease.

Stimulated by these findings, this section attempts to link learning and diffusion of technology.

5.3.1 Learning Coefficient Function Incorporating Functionality Decrease

Based on the analyses in Sect. 5.2, (5.10) can be depicted as follows by incorporating an additional term $(a_h e^{b_h t^2})$ reflecting functionality decrease in long run[2], and this should be equivalent to (5.14) over the time:

$$\lambda = \varphi_1 - \varphi_2 \left(a e^{-bt} + \frac{a_k}{1 - b_k/b} e^{-b_k t} + a_h e^{b_h t^2} \right) \tag{5.10'}$$

$$= \alpha - \beta e^{-(l-mt)t} \tag{5.14}$$

where a_h, and b_h, coefficients reflecting functionality decrease.

Comparing (5.10') and (5.14), the following conditions can be obtained within a certain period[3]:

$$\alpha = \varphi_1 + \varepsilon_1, \tag{5.16}$$

$$\varphi_2 a e^{-bt} + \frac{\varphi_2 a_k}{1 - b_k/b} e^{-b_k t} + \varphi_2 a_h e^{b_h t^2} = \beta e^{-(lt - mt^2)} + \varepsilon_2, \tag{5.17}$$

$$W(t) = \varphi_2 \cdot J(t) + \varphi_2 \cdot a_h e^{b_h t^2}, \tag{5.18}$$

where

$$W(t) \equiv \beta e^{-(l-mt)t}, \tag{5.19}$$

$$J(t) \equiv a e^{-bt} + \frac{a_k}{1 - b_k/b} e^{-b_k t}. \tag{5.20}$$

Since α and $W(t)$ are identified by (5.15), and $J(t)$ can be identified by LFDCC enumerated by (5.8), ϕ_1 as well as ϕ_2, a_h, and b_h can be identified by (5.16) and (5.17), respectively. Following these steps and applying the data obtained from empirical analysis on Japan's PV development trajectory as well as learning coefficient

---

[2] This term is equivalent to $-mt$ in (5.13). Given the small value of the power of exponent, the second term of (5.10') can be approximated as follows:

$$\begin{aligned} &-\phi_2 \left[a(1-bt) + \tfrac{a_k}{1-b_k/b}(1-b_k) + a_h \left(1 + b_h t^2\right) \right] \\ &= - \left[\left(a + \tfrac{a_k}{1-b_k/b} + b_h \right) \phi_2 - \left(ab + \tfrac{a_k b_k}{1-b_k/b} \right) \phi_2 t + a_h b_h \phi_2 t^2 \right] \\ &\equiv -(\alpha_1 - \beta_1 t + \gamma_1 t^2) \end{aligned} \tag{A}$$

While the second term of (5.14) can be approximated as follows:

$$-\beta\,[1 - (l - mt)t] = -\beta(1 - lt + mt^2) \equiv -(\alpha_2 - \beta_2 t + \gamma_2 t^2) \tag{B}$$

Under the condition within certain period, $\alpha_1 = \alpha_2$, $\beta_1 = \beta_2$ and $\gamma_1 = \gamma_2$, the structure of the additional term $a_h e^{b_h t^2}$ can satisfy the requirement of (A) and (B) are equivalent.

[3] These conditions can be satisfied and the requirements can be met after a certain period.

Table 5.3 Estimation results for the development trajectory of Japan's PV (1976–2000)

K_K	a	b	a_K	b_K	adj.R^2	DW
9.453×10^3	1.796×10^4	5.870×10^{-1}	9.472×10^2	1.670×10^{-1}	0.998	0.64
(3.31)	(1.57)	(5.74)	(3.89)	(29.79)		

trajectory in Sect. 5.2, *LFDCC driven learning coefficient function incorporating functionality decrease effects (LFDCC-LCFDE)* as depicted in (5.10') is identified.

Coefficients a, b, a_k and b_k as well as ultimate carrying capacity K_k for Japan's PV development trajectory over the period 1976–2000 are estimated in Table 5.3.

Coefficients ϕ_1, ϕ_2, a_{h} and b_{h} governing LFDCC-LCFDE for Japan's PV development are also estimated by the following approach.

First, by means of a preparatory regression using Shazam over the period 1976–2000 aiming at identifying asymptotes of the additional term $(a_h e^{b_h t^2})$ reflecting functionality decrease in long run, it was confirmed that

$$\frac{a_h}{J(t)} \ll 1 \tag{5.18a}$$

$$b_h t^2 \ll 1 \tag{5.18b}$$

From (5.18), and taking approximation based on (5.18a) and (5.18b):

$$\begin{aligned} \ln W(t) &= \ln \varphi_2 \left[J(t) + a_h e^{b_h t^2}\right] = \ln \varphi_2 \cdot J(t) \left[1 + \tfrac{a_h}{J(t)} e^{b_h t^2}\right] \\ &\approx \ln \varphi_2 + \ln J(t) + \tfrac{a_h}{J(t)} e^{b_h t^2} \approx \ln \varphi_2 + \ln J(t) + \tfrac{a_h}{J(t)}(1 + b_h t^2) \\ &= \ln \varphi_2 + \left(\ln J(t) + \tfrac{a_h}{J(t)}\right) + a_h . b_h \tfrac{t^2}{J(t)}. \end{aligned} \tag{5.18c}$$

$\ln J(t)$ can be approximated as follows[4]:

$$\ln J(t) = \eta_1 - \eta_2 \frac{1}{J(t)}, \tag{5.18d}$$

where η_1 and η_2, coefficients.

Substituting $\frac{1}{J(t)}$ in (5.18d) for $\frac{1}{J(t)}$ in (5.18c)

$$\ln W(t) = \left(\ln \varphi_2 + a_h \cdot \frac{\eta_1}{\eta_2}\right) + \left(1 - \frac{a_h}{\eta_2}\right) \ln J(t) + a_h \cdot b_h \frac{t^2}{J(t)}, \tag{5.18e}$$

[4] Equation (5.11) suggests that under certain conditions, $J(t)$ can be approximated as follows:

$$\begin{aligned} &J(t) \approx \beta e^{-\gamma t} (0 < \beta,\ 0 < \gamma \ll 1) \\ &\ln J(t) = \ln \beta - \gamma t = (\ln \beta + 1) - (1 + \gamma t) \\ &\approx (1 + \ln \beta) - e^{\gamma t} = (1 + \ln \beta) - \frac{\beta}{J(t)} \equiv \eta_1 - \eta_2 \frac{1}{J(t)} \end{aligned}$$

$$\equiv \omega_1 + \omega_2 \ \ln J(t) + \omega_3 \cdot \frac{t^2}{J(t)}, \tag{5.18f}$$

$$\text{where } \omega_1 = \ln \varphi_2 + a_h \cdot \frac{\eta_1}{\eta_2}, \tag{5.18g}$$

$$\omega_2 = 1 - \frac{a_h}{\eta_2}, \tag{5.18h}$$

$$\omega_3 = a_h \cdot b_h. \tag{5.18i}$$

From (5.18h):

$$a_h = (1 - \omega_2)\,\eta_2. \tag{5.18j}$$

From (5.18i):

$$b_h = \frac{\omega_3}{a_h} = \frac{\omega}{(1-\omega_2)\eta_2}. \tag{5.18k}$$

From (5.18g):

$$\begin{aligned} \ln \varphi_2 &= \omega_1 - a_h \cdot \frac{\eta_1}{\eta_2} = \omega_1 - (1-\omega_2)\,\eta_1, \\ \varphi_2 &= Exp\ (\omega_1 - (1-\omega_2)\,\eta_1). \end{aligned} \tag{5.18l}$$

By means of regression analyses over the period 1981–2000 utilizing data obtained from Tables 5.1 and 5.3, the following results were obtained:

$$\ln W(t) = -4.938 + 2.095 \times 10^{-2} \ \ln J(t) + \ 2.190 \times 10^{-4} \frac{t^2}{J(t)} - 2.811 \times 10^{-3} D,$$

$$\begin{array}{llll} & & \text{adj.R}^2 & \text{DW} \\ (-2{,}570.72)\ (60.45)\ (3.86)\ (-3.93) & & 0.998 & 1.59, \end{array} \tag{5.18m}$$

where D, dummy variables (1981, 1982, 1995 = 1, other years = 0).

$$\begin{array}{lll} \ln J(t) = 8.360 - 1.177 \dfrac{1}{J(t)} & \text{adj. R}^2 & \text{DW} \\ \qquad\qquad (18.52)\ (-12.26) & 0.931 & 1.33. \end{array} \tag{5.18n}$$

(By means of Cochrun–Orcutt treatment).

From (5.18m) and (5.18n) coefficients a_h, b_h, and ϕ_2 were identified as follows:

$$\begin{aligned} a_h &= \left(1 - 2.095 \times 10^{-2}\right) \times 1.177 = 1.152, \\ b_h &= 2.190 \times 10^{-4}/1.152 = 1.901 \times 10^{-4} \\ \varphi_2 &= Exp\left(-4.938 - \left(1 - 2.095 \times 10^{-2}\right) \times 8.360\right) = Exp\,(-13.122) = 2.001 \times 10^{-6}. \end{aligned}$$

By applying identified a_h, b_h, and ϕ_2 to (5.10') and taking balance with (5.14), ϕ_1 $(= \alpha + \varepsilon_1)$ was estimated as:

$$\varphi_1 = 3.478 \times 10^{-1}.$$

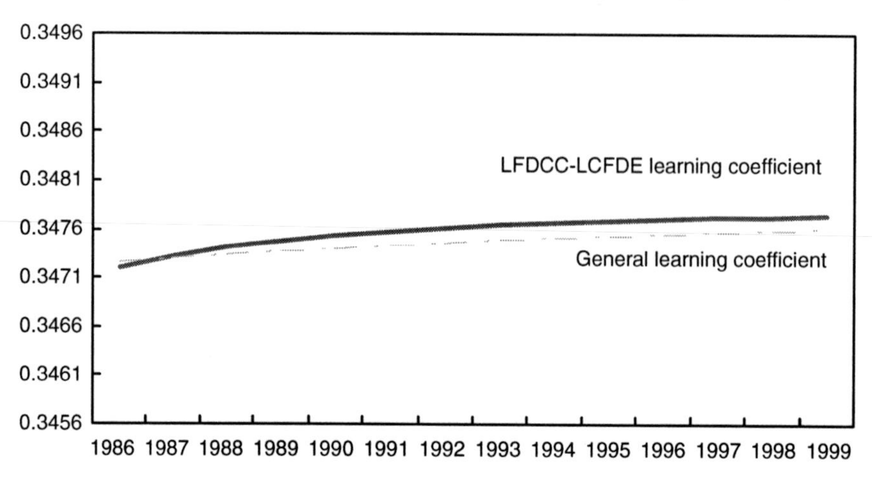

Fig. 5.5 Trends in learning coefficient in Japan's PV development (1986–1999)

Therefore, LFDCC-LCFDE is enumerated as follows:

$$\lambda = 3.478 \times 10^{-1} - 2.001 \times 10^{-6} \left(1.796 \times 10^{4}\, e^{-5.870\times 10^{-1} t} + 1.324 \times 10^{3}\, e^{-1.67\times 10^{-1} t} + 1.152\, e^{1.901\times 10^{-4} t^{2}}\right). \quad (5.21)$$

Utilizing estimated LFDCC-LCFDE, trend in learning coefficient in Japan's PV development over the period 1986–1999 is illustrated in Fig. 5.5 by comparing the trend in learning coefficient measured by the learning coefficient function considering functionality decrease as illustrated in Fig. 5.3. The estimate by LFDCC-LCFDE also corresponds to the learning coefficient of Japan's leading PV firms as demonstrated in Table 5.2. Figure 5.5 shows that estimate by LFDCC-LCFDE demonstrates convex behavior of learning coefficient more clearly.

This demonstrates the reliability of (a) the general learning coefficient function considering functionality decrease effects (5.14), and (b) the mathematical structure of the factor reflecting functionality decrease in long run ($a_h e^{b_h t^2}$ in (5.10')).

5.3.2 Technology Diffusion Trajectory Reflecting Functionality Decrease Effects

The series of the analyses in Sect. 5.3.1 demonstrate that functionality decrease effects on learning coefficient inevitably affect the trajectory of technology diffusion in long run which compels a modification in the logistic growth function within a dynamic carrying capacity (LFDCC) depicted by (5.8). Furthermore, mathematical development process from (5.6) to (5.10) together with (5.10') suggest that LFDCC could reflect functionality decrease effects in long run by adding an additional term $(a_h e^{b_h t^2})$, which was demonstrated as reflecting functionality decrease in long run, in its denominator as follows:

$$Y^*(t) = \frac{K_k}{1 + a\, e^{-bt} + \frac{a_k}{1-b_k/b} e^{-b_k t} + a_h e^{b_h t^2}}. \tag{5.22}$$

Equation (5.22) suggests that diffusion trajectory would be depressed in long run by the functionality decrease term. Given the LFDCC incorporating functionality decrease effects (LFDCC-FDE) as enumerated by (5.22), its dynamic carrying capacity is enumerated as follows (see Appendix for mathematical development):

$$K(t) = \frac{K_k}{1 + a_k e^{-b_k t} + [b(b + 2b_h t)]\, a_h e^{b_h t^2}}. \tag{5.23}$$

Equation (5.23) suggests that the impacts of the additional term derived from functionality decrease effects reveal significantly in depressing carrying capacity as time runs by in long run.

Given the mechanism in creating a new carrying capacity in the process of IT diffusion as illustrated in Fig. 5.6[5], decrease in carrying capacity reacts to decrease in diffusion, which again decreases carrying capacity resulting in a vicious cycle between stagnation of diffusion and carrying capacity as illustrated in Fig. 5.7. At the same time, decrease in carrying capacity accelerates obsolescence of technology.

Figure 5.7 suggests that systems restructuring is indispensable for virtuous cycle, and activation of interaction with institutional systems plays a significant role for this restructuring.

Figure 5.8 and Table 5.4 demonstrate trajectory of Japan's PV development measured by (5.22) comparing with that of estimated by LFDCC as well as the actual trend.

Figure 5.8 and Table 5.4 demonstrate that LFDCC-FDE estimate demonstrates slightly closer to actual trend than that of estimate by LFDCC as time runs by. In addition, they demonstrate that as far as the estimate by 2000 is concerned, there are no substantial impacts with respect to the foregoing vicious cycle between stagnation of carrying capacity and diffusion derived from the functionality decrease effects in learning.

However, Fig. 5.9, which estimates the future trajectory of dynamic carrying capacity in long run by using (5.23), indicates that in long run carrying capacity

[5] In the process of IT diffusion, the number of users increases as time passes, which induces interaction with institutions leading to increasing potential users by increased value and function as the network externalities gain momentum. Thus, IT creates new demand in this development process and new functionality is formed which in turn enhances user interaction. Thus, the interactive self-propagating behavior continues [20, 21].

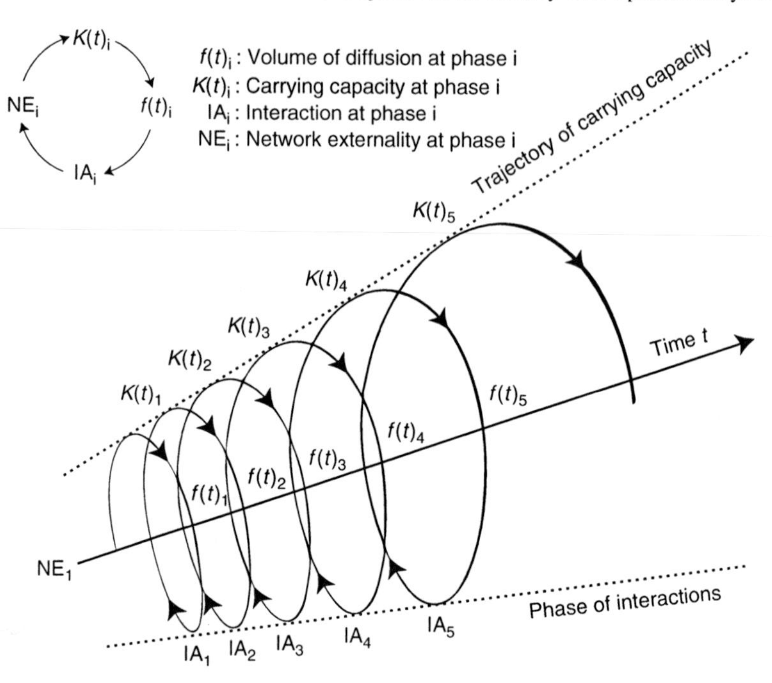

Fig. 5.6 Mechanism in creating a new carrying capacity in the process of IT diffusion Original source: Watanabe et al., 2001 [22]

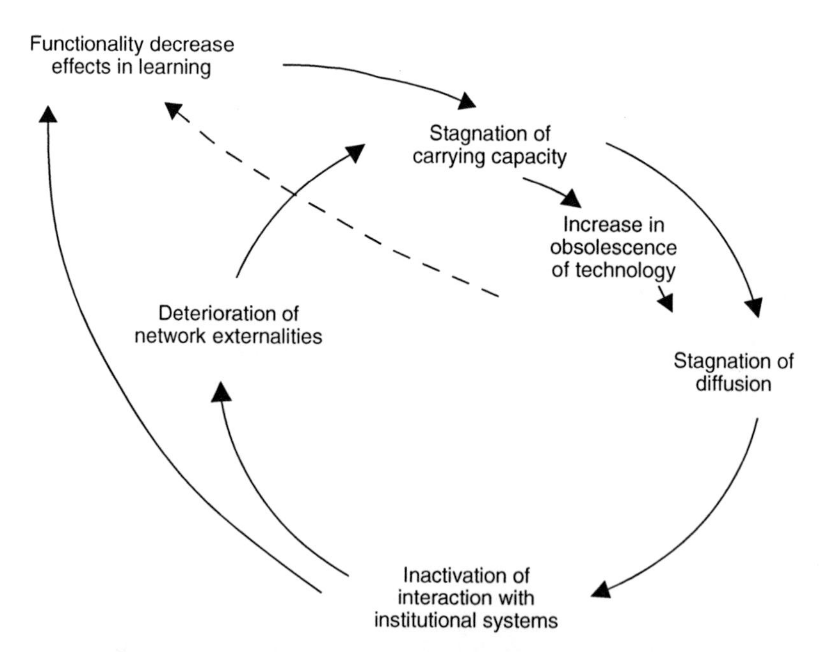

Fig. 5.7 Impacts of functionality decrease in learning leading to a vicious cycle between diffusion and carrying capacity

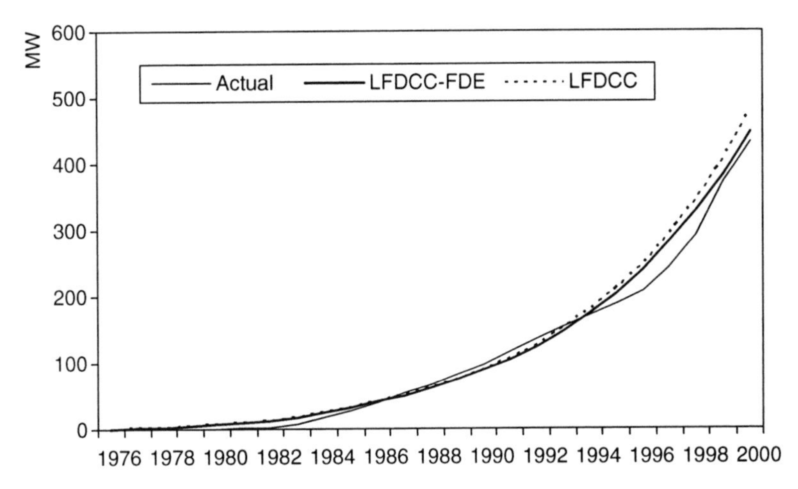

Fig. 5.8 Diffusion trajectory of Japan's PV development (1976–2000): MW

Table 5.4 Comparison of diffusion trajectory estimates in Japan's PV development (1986–2000): MW

Year	Actual	LFDCC-FDE	LFDCC
1986	40.43	41.12	41.34
1987	53.63	50.52	50.86
1988	66.43	61.28	61.77
1989	80.63	73.66	74.38
1990	97.43	87.98	89.01
1991	117.23	104.62	106.10
1992	136.03	124.00	126.10
1993	152.73	146.60	149.56
1994	169.23	172.95	177.11
1995	186.63	203.65	209.48
1996	207.83	239.33	247.50
1997	242.83	280.71	292.12
1998	291.83	328.52	344.41
1999	371.83	383.55	405.60
2000	432.76	446.56	477.05

will be dramatically stagnated by functionality decrease effects which provides a significant fear to a vicious cycle as illustrated in Fig. 5.7. Figure 5.9 indicates that this stagnation becomes distinct after the year 2009 corresponding to the year when learning coefficient changes to decrease as estimated in Fig. 5.4.

All support the significance of incorporating functionality decrease effects in estimating both learning coefficients and diffusion trajectory of innovative goods. In addition, the foregoing analysis demonstrates the significance of the interaction between learning and diffusion of technology.

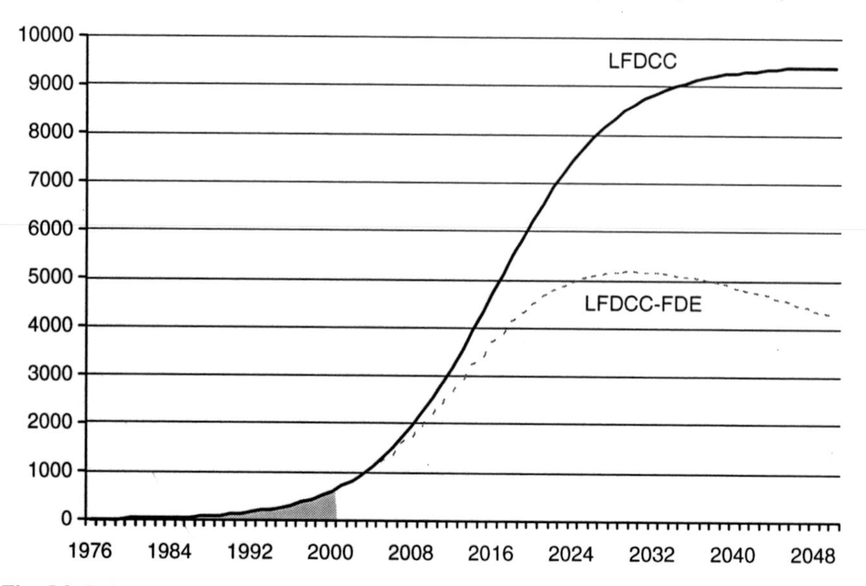

Fig. 5.9 Estimates of the trajectory of dynamic carrying capacity in Japan's PV development (1976–2050): MW

5.3.3 *Linking Learning and Diffusion of Technology*

On the basis of the foregoing mathematical analysis and empirical demonstration, interaction between learning and diffusion of technology is identified leading to systematic measurement of (a) LFDCC (logistic growth function within a dynamic carrying capacity) based learning coefficient incorporating functionality decrease effects (LFDCC-LCFDE) and (b) LFDCC incorporating functionality decrease effects (LFDCC-FDE) as illustrated in Fig. 5.10. Table 5.5 summarizes an algorithm for this stepwise systematic measurement.

5.4 Institutional Dynamism Leading to a Dynamic Interaction Between Learning, Diffusion and Spillover of Technology

The analysis in Sect. 5.3 demonstrates the significance of the interaction between learning and diffusion of technology. This interaction induces vigorous R&D activities which lead to increasing technology stock (T). Technology stock, in turn, as a direct result of R&D investment (R) inevitably stimulates multi-factor learning (*MFL*) [18, 19]. Multi-factor learning induces further increase in technology stock. This necessitates both indigenous R&D investment and effective utilization of spillover technology (T_s).

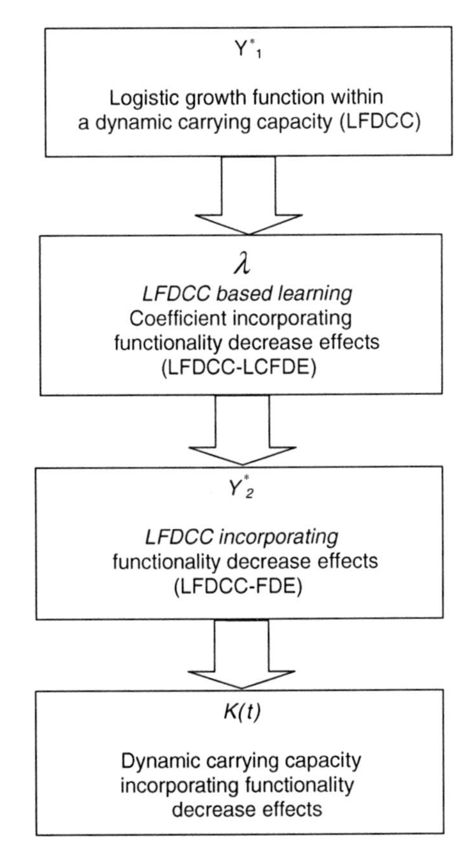

Fig. 5.10 Steps for measurement of LFDCC-LCFDE and LFDCC-FDE

Trans-generational technology spillovers accumulates learning, and learning can be considered as one of the sources of spillovers as well as being considered as an effect of spillovers at the same time. Learning and spillovers together with technology stock generated by indigenous R&D enhance total factor productivity (*TFP*), as illustrated in Fig. 5.11, which in turn contributes to production increase (Y). Increased production results in higher cumulative production (Y^*) which stimulates learning. Furthermore, it induces R&D investment, which in turn generates technology stock. Thus, an organic comprehensive structure led by institutional dynamism generating dynamic interaction between learning, diffusion and spillover of technology is constructed.

A scheme in generating this dynamism is summarized in Table 5.6.

Figure 5.12 illustrates an institutional dynamism leading to the foregoing dynamic interaction between learning, diffusion and spillover of technology.

As analyzed in Sect. 5.3, systems restructuring is indispensable for shifting a vicious cycle between stagnation of diffusion and carrying capacity, and activation of interaction with institutional systems play a significant role for this restructuring.

Table 5.5 The algorithm for systematic measurement of learning coefficient and diffusion trajectory based on a logistic growth function within a dynamic carrying capacity and incorporating functionality decrease effects (LFDCC-LCFDE and LFDCC-FDE)

First step	Estimate a trajectory of Y^* by means of a logistic growth function within a dynamic carrying capacity (LFDCC) approach $Y_1^* = \frac{K_k}{1+ae^{-bt}+\frac{a_k}{1-b_k/b}e^{-b_k t}}$ (1)
Second step	Estimate a general learning coefficient function considering functionality decrease effects $\lambda_1 = \alpha - \beta e^{(l-mt)t}$ $(\lambda_{1\,\max} = \alpha - \beta e^{\frac{l^2}{4m}}$ at $t = \frac{l}{2m}$ (when $\frac{d\lambda}{dt} = 0$)) (2)
Third step	Estimate an adjusted LFDCC based learning coefficient λ by introducing a term reflecting functionality decrease effects $\lambda_2 = \varphi_1 - \varphi_2\left(ae^{-bt} + \frac{a_k}{1-b_k/b}e^{-b_k t} + a_h e^{b_h t^2}\right)$ (3)
Fourth step	Identify LFDCC based learning coefficient incorporating functionality decrease effects (LFDCC-LCFDE) $\lambda_3 = \alpha - \beta e^{-(l-mt)t} = \varphi_1 - \varphi_2\left(ae^{-bt} + \frac{a_k}{1-b_k/b}e^{-b_k t} + a_h e^{b_h t^2}\right)$ (4) where $\begin{pmatrix} W(t) = \phi_2 \cdot J(t) + \phi_2 \cdot a_h e^{b_h t^2} \\ W(t) \equiv \beta e^{-(l-mt)t} \text{ and } J(t) \equiv ae^{-bt} + \frac{a_k}{1-b_k/b}e^{-b_k t} \end{pmatrix}$
Fifth step	Identify LFDCC incorporating functionality decrease effects (LFDCC-FDE) $Y^* = \frac{K_k}{1+ae^{-bt}+\frac{a_k}{1-b_k/b}e^{-b_k t}+a_h e^{b_h t^2}}$ (5)
Sixth step	Identify dynamic carrying capacity for LFDCC-FDE $K(t) = \frac{K_k}{1+a_k e^{-b_k t}+[b(b+2b_h t)]a_h e^{b_h t^2}}$ (6)

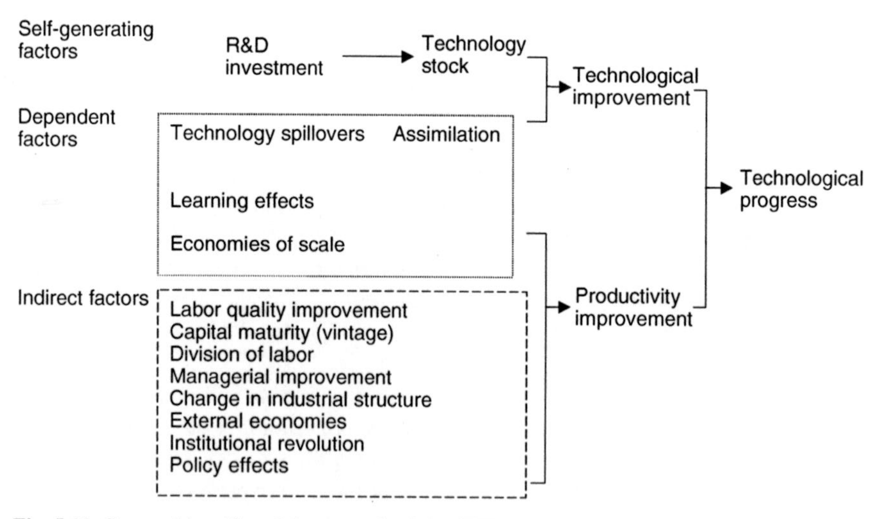

Fig. 5.11 Composition of total factor productivity (TFP)

Table 5.6 Scheme in generating dynamism between learning, diffusion and spillover of technology

(a) Technology stock as a direct result of R&D investment inevitably stimulates multifactor learning (MFL)
(b) MFL induces technological progress (TP)
(c) Technological progress (TP) necessitates both indigenous R&D (R) and resulting T_i and effective utilization of spillover technology (T_s)
(d) Trans-generational spillovers accumulates learning
(e) Learning can be considered as one of the sources of spillovers as well as being considered as an effect of spillovers
(f) Learning, spillovers together with technology stock generated by R&D enhance TFP
(g) Enhanced TFP contributes to production increase (Y)
(h) Increased production leads to higher cumulative production (Y^*) which stimulates learning, in addition, it induces R&D investment which in turn generates technology stock

Figure 5.12 supports these postulate and suggests that the significant of institutional elasticity for activating interaction with institutional systems leading to a positive dynamic interaction between learning, diffusion and spillovers of technology.

5.5 Conclusion

In light of the increasing significance of the systems approach in maximizing the effects of innovation by means of the effective utilization of the potential resources of innovation, this chapter undertook theoretical analysis of this subject focusing on a dynamism between learning and diffusion of technology. An empirical demonstration was also attempted taking Japan's PV development trajectory, which follows the similar trajectory of IT's functionality development, over the last quarter century.

Based on these analyses, dynamism between learning and diffusion of technology was elucidated, thereby the effects of functionality decrease on learning coefficient and consequent impacts on technology diffusion and its carrying capacity were identified.

Noteworthy findings include:

(a) On the basis of intensive empirical analyses and reviews of proceeding works, it was anticipated that the behavior of learning coefficient has close relevance with that of a logistic growth function within a dynamic carrying capacity. This coefficient was anticipated to increase as a consequence of cumulative learning effects and change to decreasing trend in long run as functionality decreases.

(b) Such a dynamic convex behavior of learning coefficient was enumerated by an equation derived from a logistic growth function within a dynamic carrying capacity with an additional term reflecting functionality decrease in long run. On the basis of an empirical analysis by applying this equation in Japan's PV development trajectory over the last quarter century, it was demonstrated that this equation reflected the learning coefficient of Japan's PV firms, thereby

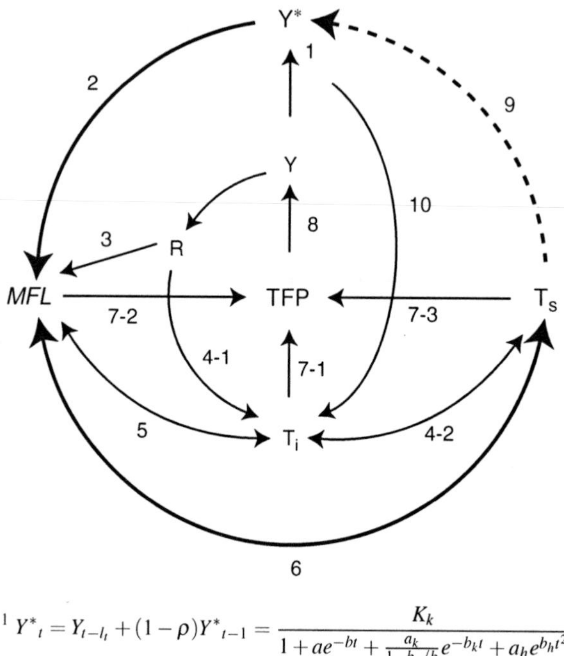

1 $Y^*_t = Y_{t-l_t} + (1-\rho)Y^*_{t-1} = \dfrac{K_k}{1+ae^{-bt}+\frac{a_k}{1-b_k/b}e^{-b_k t}+a_h e^{b_h t^2}}$

2 $\lambda = \alpha - \beta e^{-(l-mt)t} = \phi_1 - \phi_2(ae^{-bt} + \dfrac{a_k}{1-b_k/b}e^{-b_k t} + a_h e^{b_h t^2})$

3 $MFL = f(\lambda) = f(\lambda(Y^*, T(R))$

4

4-1 $T_{it} = R_{t-l_t} + (1-\rho)T_{it-1}$

4-2 $T = T_\mathrm{i} + Z \cdot T_\mathrm{s}$

5 $MFL = g(\lambda) = g(\lambda(Y^*,\ T(T_{i,}\ (R),\ ZT_s)))$

6 $MFL = h(Y^*,\ \int_0^t T_\mathrm{s}\,\mathrm{d}s)$

$T_\mathrm{s} = T_\mathrm{s}(MFL)$

7 $TFP = TFP(T_\mathrm{i},\ MFL,\ T_\mathrm{s})$

8 $Y = F(L,\ K,\ TFP)$

9 $Y^*_t = Y_{t-l_t}(L,\ K,\ TFP(T_\mathrm{i},\ MFL,\ T_\mathrm{s})) + (1-\rho)Y^*_{t-1}$

10 $\rho_t = \rho(K(t)) = \rho(K_t/1 + a_k e^{-b_k t} + a_h b(2b_h t + b)e^{b_h t^2})$

MFL, multifactor learning; TFP, total factor productivity; L, labor; K, capital; Y, production; Y^*, cumulative production; R, R&D investment; T, technology stock; T_i, indigenous technology; T_s, spillover technology; Z, assimilation capacity; ρ, rate of obsolescence of technology

Fig. 5.12 Institutional dynamism leading to a dynamic interaction between learning, diffusion and spillover of technology

the significance of this equation was demonstrated. This dynamic coefficient function incorporating functionality decrease effects revealed that an estimate without considering functionality decrease effects leads to higher estimate than that of estimated by reflecting functionality decrease effects.

(c) Synchronizing this equation in a logistic growth function within a dynamic carrying capacity, an equation depicting diffusion trajectory of innovative goods incorporating functionality decrease effects was developed which demonstrates similar trajectory as actual one, thereby significance of this equation was demonstrated. A trajectory estimated by this equation demonstrates slightly lower diffusion trajectory than the trajectory estimated by a normal logistic growth function within a dynamic carrying capacity without considering functionality decrease effects. This was considered due to a "depression effect" as a consequent of functionality decrease.

(d) Based on this new logistic growth function within a dynamic carrying capacity incorporating functionality decrease effects, the impacts of this functionality decrease on the dynamic carrying capacity was analyzed. The analysis identified that this impact is not so significant in short term, significant impact in long run in stagnating carrying capacity as time runs by was revealed. In addition, it was identified that the decrease in this carrying capacity accelerates obsolescence of technology. This significant impact was identified to lead a vicious cycle between stagnating carrying capacity and diffusion trajectory.

Important suggestions supportive to nations technology policy and firms R&D strategy in light of the maximum utilization of the potential resources for innovation under a long lasting economic stagnation while facing a new paradigm initiated by an information society can be focused on the following points:

(a) Systems restructuring is indispensable for shifting a vicious cycle between stagnation of diffusion and carrying capacity. Given the IT's self-propagating nature formation process in which interaction with institutions plays a significant role, activation of interaction with institutional systems plays a significant role to this restructuring.

(b) In this consequence, a way to lead a positive dynamic interaction between learning, diffusion and spillovers of technology depends on institutional elasticity for activating interaction with institutional systems.

Given that the state of institutional systems constructs a virtuous cycle between techno-economic development of the nation, shifting current vicious cycle to a virtuous cycle would be crucial.

Points of further works are summarized as follows:

(a) Further elaboration of the relationship between the state of institutional system, more specifically institutional elasticity, the state of innovation and diffusion and trend in functionality.

(b) International comparison of the institutional elasticity and its effect on innovation and diffusion of technology.

(c) Demonstration of the significance of institutional elasticity and its contribution in maximizing the effects of policy.

References

1. OECD, Technology Productivity and Job Creation (OECD, Paris, 1998)
2. C. Watanabe, The feedback loop between technology and economic development: an examination of Japanese industry, Technological Forecasting and Social Change 49, No. 2 (1995) 127–145
3. C. Watanabe, The perspective of techno-metabolism and its insight into national strategies, Research Evaluation 6, No. 2 (1997) 69–76
4. C. Watanabe, R. Kondo, N. Ouchi, H. Wei, Formation of IT features through interaction with institutional systems – empirical evidence of unique epidemic behavior, Technovation 23, No. 3 (2003) 205–219
5. V.W. Ruttan, Technology, Growth, and Development – An Induced Innovation Perspective (Oxford University Press, New York, 2001) 118–137
6. C. Watanabe, C. Griffy-Brown, B. Zhu and A. Nagamatsu, Inter-firm technology spillover and the 'virtuous cycle': photovoltaic development in Japan, in A. Gruebler, N. Nakicenovic and W.D. Nordhaus (eds.). Technological Change and the Environment (Resources for the Future, Washington DC, 2001)
7. C. Watanabe, B. Asgari and A. Nagamatsu, Virtuous cycle between R&D, functionality development and assimilation capacity for competitive strategy in Japan's high-technology industry, Technovation 23, No. 11 (2003) 879–900
8. D.S. Price, Little Science, Big Science (Columbia University Press, New York, 1965)
9. W.M. Cohen and D.A. Levinthal, Absorptive capacity: a new perspective of learning and innovation, Administrative Science Quarterly 35, No. 1 (1990) 128–153
10. K. Arrow, The economic implications of learning by doing, Review of Economic Studies 29 (1962) 155–173
11. N. Rosenberg, Factors affecting the diffusion of technology, in his book Perspectives on Technology (Cambridge University Press, Cambridge, 1976) 189–210
12. E.M. Rogers, Diffusion of innovation (The Free Press of Glencoe, New York, 1962)
13. J.S. Metcalfe, The diffusion of innovation in the lancashire textile industry, Manchester School of Economics and Social Studies 2 (1970) 145–162
14. J.S. Metcalfe, Impulse and diffusion in the study of technical change, Futures 13, No. 5 (1981) 347–359
15. P.S. Meyer, Bi-logistic growth, Technological Forecasting and Social Change 47, No. 1 (1994) 89–102
16. P.S. Meyer, and J.H. Ausbel, Carrying capacity: a model with logistically varying limits, Technological Forecasting and Social Change 61, No. 3 (1999) 209–214
17. IEA, Experience Curves for Energy Technology Policy (OECD/IEA, Paris, 2000)
18. W.M. Cohen and D.A. Levinthal, Innovation and learning: the two faces of R&D, The Economic Journal 99 (1989) 569–596
19. N. Kouvaritakis, A. Soria and S. Isoard, Modeling energy technology dynamics: methodology for adaptive models with learning by doing and learning by searching, International Journal of Global Energy Issues 14 (2000) 104–115
20. R.R. Nelson and B.N. Sampat, Making sense of institutions as a factor shaping economic performance, Journal of Economic Behavior & Organization 44 (2001) 31–54
21. D.C. North, Institutions, institutional change, and economic performance through time, The American Economic Review 84 (1994) 359–368
22. C. Watanabe, Systems option for sustainable development, in OECD (ed.), Energy the Next Fifty Years (OECD, Paris,1999) 121–146

Chapter 6
Diffusion, Substitution and Competition Dynamism Inside the ICT Market: A Case of Japan

Abstract Under the new information society paradigm that emerged in the 1990s, contrary to its conspicuous achievement as an industrial society, Japan is experiencing a vicious cycle between non-elastic institutions and insufficient utilization of the potential benefits of information and communication technology (ICT).

However, a dramatic deployment of mobile telephones with internet access service such as NTT DoCoMo's i-mode service in the late 1990s provides encouragement that, once the potential is exploited, Japan's institutional systems can effectively stimulate the self-propagating nature of ICT. The rapid deployment of internet protocol (IP) mobile service in Japan can be attributed to worldwide advances in the utilization of personal computers (PCs) and the internet. Thus, a complex technology web triggered by the dramatic advancement of PCs and the internet and co-evolving diffusion, substitution and competition dynamism has emerged in the global ICT market, particularly in Japan's mobile communication business.

The above observations prompt the hypothetical view that, despite a lack of institutional elasticity, recent advances in Japan's IP mobile service deployment can be attributed to a co-evolutionary dynamism between diffusion, substitution and competition inside the ICT market. Thus, policy questions could be how to create such a co-evolutionary dynamism by means of ICT innovation, enriched functions, reduced price and competitive environment.

In order to demonstrate the foregoing hypothesis, an empirical analysis of the mechanism co-evolving diffusion, substitution and competition dynamism inside Japan's ICT market is attempted by utilizing four types of diffusion models identical to respective diffusion dynamics.

Reprinted from *Technological Forecasting and Social Change* 73, No. 6, C. Chen and C. Watanabe, Diffusion, Substitution and Competition Dynamism Inside the ICT Market: A Case of Japan, pages: 731–759, copyright (2006), with permission from Elsevier.

C. Watanabe, *Managing Innovation in Japan: The Role Institutions Play in Helping or Hindering how Companies Develop Technology*, DOI: 10.1007/978-3-540-89272-4_6,

6.1 Introduction

Under the new paradigm of an information society which emerged in the 1990s, contrary to its conspicuous achievement in an industrial society, Japan is experiencing a vicious cycle between non-elastic institutions and insufficient utilization of the potential benefits of information and communication technology (ICT or IT, hereinafter refers as ICT) [1].

Comparison of the growth rate of ICT investment in G7 countries over the period 1985–1996 demonstrates that Japan's ICT investment level ranks the third after the US and UK, and is comparable to other G7 countries [2].

Notwithstanding such investment in ICT, Japan's utilization of its benefits is hardly satisfactory as pointed out by OECD in 2001 [3]. OECD demonstrates the relationship between the access costs and a diffusion of the internet in OECD countries which reveals that countries with lower access costs have more internet hosts. However, in spite of its moderate internet access cost enabled by the foregoing ICT investment, Japan does not achieve the expected internet penetration rate. By contrast, the US performs fairly well. This gap should be attributed to the contrastive institutional elasticity of the two countries in an information society [1].

However, a dramatic deployment of mobile telephone with internet access service (internet protocol (IP) mobile) such as i-mode service (NTT DoCoMO's mobile internet access service) in the late 1990s provides encouragement that, once the potential is exploited, Japan's institution systems can effectively stimulate the self-propagating nature of ICT through dynamic interaction with them as is typically observed in high level utilization of IP mobile. ITU demonstrates that Japan's dependency on IP mobile out of total mobile telephony is conspicuously high as 72.3% in 2001 while the same ratio in the US and Germany is 7.9% [4].

Such a rapid deployment in IP mobile in Japan can be attributed to a worldwide dramatic advancement of personal computers (PCs) and the internet [5].

In addition, rapidly developing trans-generationary substitution of mobile communicates system accelerates such dramatic IP mobile deployment [6].

Thus, a *complex technology web triggered by the dramatic advancement of PCs and the internet and co-evolving diffusion, substitution and competition dynamism* as illustrated in Fig. 6.1 emerges in the world wide ICT market, particularly in Japan's mobile communication business.

These observations prompt us a hypothetical view that a dramatic deployment of Japan's IP mobile despite its institutional less elasticity can be attributed to a *co-evolutionary dynamism between diffusion, substitution and competition* inside its ICT market. Thus, policy questions could be how to create such a co-evolutionary dynamism by means of ICT innovation, enriched functions, reduced price and competitive environment.

Early applications of *diffusion models* addressed primarily durable goods market such as TV set, refrigerator, etc. The well-known first-purchase diffusion new products in marketing are those of Woodlock and Fourt [8], Mansfield [9], Bass [10], Easingwood, Mahajan and Muller [11], etc. While Bass took diffusion speed coefficient as a constant, the nonsymmetric responding logistic (NSRL) model,

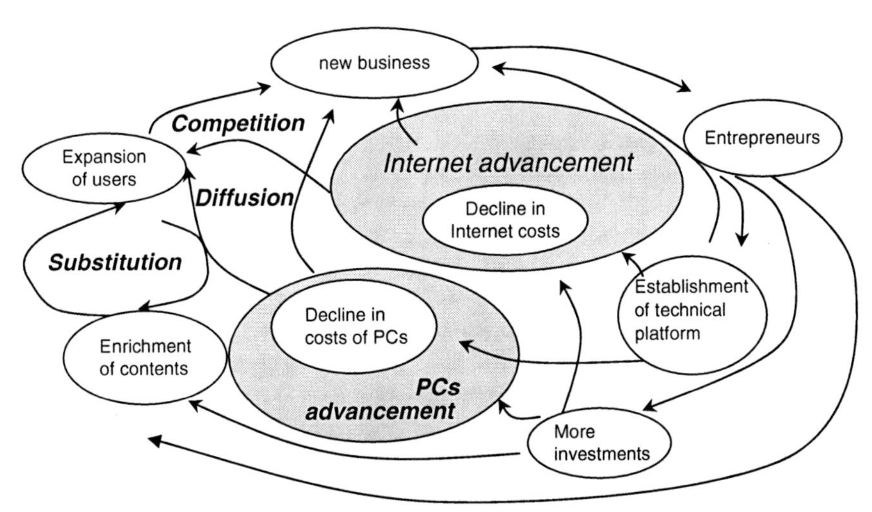

Fig. 6.1 Complex technology web emerging Japan's mobile communication business
Source: Authors' elaboration based on [7]

proposed by Easingwood et al., treated the diffusion speed coefficient as a function of multiplication of diffusion rate so that it decreases or increases depending on the situations. These models were *single-product models* concerned only with the sales growth for a single product. The often-criticized insufficiency of them is that new products or technologies are not adopted into a completely new market without any other existing similar or related technology. The existence of other products and technologies may affect, no matter positively or negatively, the sales of a new product.

Although those first-purchase models succeeded in illuminating the curve of diffusion at first, due to the increasing diffusion potential of technologies resulted from the increasing population, existence of other products or upgraded function of technology, constant potential has become the conceptual limit. In order to overcome the restriction and single-product models, Norton and Bass [12], Mahajan and Muller [13] and Jun and Park [14] proposed *multi-generation diffusion models* that simultaneously trace *the diffusion and substitution trends* for the successive generations of a durable technological innovation. Mahajan and Muller [13] used their model to explore the diffusion of four generations of IBM mainframe computers and signified that the introduction time of a new technology is a key element in the new product strategy. While Norton and Bass [12] and Mahajan and Muller [13] just extended the form of the Bass model, Jun and Park [14], by taking the utility function of consumers into account, proposed a diffusion model for multiple generations of products representing customer choice behavior.

Competitive influence was first discussed in the marketing literature by Peterson and Mahajan [15]. As Parker and Gatignon [16] noted, the optimal strategy for various marketing mix variables has been an integral part of most models of brand-level or competitive market behavior. These models incorporated price [17, 18], advertising [19,20], and both price and advertising [21]. In recent years, much interest has focused on developing aggregate diffusion models in a competitive environment. Mahajan, Sharma and Buzzell [22] proposed a diffusion modeling approach for assessing the impact of entry by a new competitor into a market previously served by a single firm. Parker and Gatignon [16] investigated diffusion at the brand level, i.e. in the context of products or firms that compete in a new market. Effects of each *user's history on the diffusion* were first analyzed by Allison [23], Oren and Rothkopf [24] and Hedstrom [25], etc. using multinomial logit model in a diffusion context. Aiming at representing the processes generating events in discrete-time units, Allison [23] specified a discrete-time hazard rate as a function such as logit model, which depended on time and the explanatory variables. The method was to break up each individual's event history into a set of discrete time units in which an event either did or did not occur. Hedstrom [25] also estimated the hazard rate using the logit model based on the discrete-time event history approach. Moreover, Jun and Park [14] advocated studying the utility-maximizing choice behavior of customers to understand the sales patterns of new products and to develop sales forecasting models for multiple generations of products in a competitive market.

Concerning the *substitution of technologies in the ICT market*, such as fixed phone and mobile phone, the spread of mobile services comes at a time when telecommunications authorities and the public are concerned over the lack of development of competition in local services. Cadima and Barros [26] estimated the diffusion of Portuguese fixed-line and mobile networks, and concluded that mobile telephone adoption slowed fixed-line growth, while fixed-line subscription growth had no impact on mobile subscription growth. Gruber and Verboven [27] found, for the European Union, a negative impact of fixed-line network size on mobile telephone subscription, i.e. fixed-line and mobile service are substitute. Rodini et al. [28] analyzed fixed and mobile services for telecommunications access using a large US household survey conducted over the period 2000–2001 and concluded that substitutability between fixed and mobile telephone service impacts public policy toward *competition in both of these market*. In the US, the principal concern over competition in these market derives from the market power initiated by providers of fixed-line local telephone service. However, connection quality and number portability discrepancies between the two services are fading, and substitutability may increase over time due to continued price declines and greatly improved connection quality. Furthermore, upgraded functions and improvements of mobile services will outpace those of fixed service.

While all these analyses relevant to the PCs and the internet driven substitutions of technologies in the ICT market provide supportive evidence to the aforementioned hypothetical view that a dramatic deployment of Japan's IP mobile can be attributed to a dynamism between diffusion, substitution and competition inside Japan's ICT market, they are hardly satisfactory to elucidate the mechanism of the dynamism.

In light of the foregoing, an empirical analysis of the mechanism governing diffusion, substitution and competition dynamism inside Japan's ICT market is attempted by utilizing the diffusion models identical to respective diffusion dynamics. First the

Simple Logistic Diffusion Model is used for realizing a primary idea about the market size with the diffusion potential and parameters. Next, the structural change in the market is analyzed by using the *Bi-logistic Model and Logistic Growth within a Dynamic Carrying Capacity (LGDCC) Model*, which is an extended form of Simple Logistic Function, taking the phenomenon of potential growth into consideration. Last, in order to understand the internal factors affecting consumers and the market mechanism, the substitution modeling approach is introduced based on "*Choice-based Substitution Diffusion Model*" proposed by Jun and Park [16, 29].

Section 6.2 presents the methodology of this research, including model synthesis and data construction. Section 6.3 provides the results of empirical analysis and clear description about the ICT market in Japan characterized by mobile telephony, telephony and the internet. Section 6.4 briefly summarizes the conclusion and direction for further research.

6.2 Methodology

6.2.1 Model Synthesis

6.2.1.1 Simple Logistic Model

The simple logistic model depicted in (6.1) and (6.2) is used to estimate the size and primary structure of the ICT market taking whole telephony, mobile telephony and the internet ignoring the difference of service quality or content such as fixed or mobile in telephony, IP and NonIP in mobile phone and access methods in the internet, the user of ICT can be simply defined as the service subscriber of respective ICT examined.

$$\frac{\mathrm{d}Y(t)}{\mathrm{d}t} = \left(a\frac{Y(t)}{N}\right)(N - Y(t)), \tag{6.1}$$

$$Y(t) = \frac{N}{1 + \mathrm{e}^{-at+b}}, \tag{6.2}$$

where $Y(t)$, number of users at the time t; N, constant potential of the technology (carrying capacity); a, imitation coefficient; and b, coefficient denoting the initial status of the market.

6.2.1.2 Bi-Logistic Model

The bi-logistic model is useful in modeling many systems that contain complex growth processes not well or completely described by the single logistic function such as the co-existence of IP and NonIP subscribers in the mobile phone and dial-up and broadband in the internet access. By adding two single logistic models together, the bi-logistic model can be depicted as (6.3).

$$Y(t) = Y_1(t) + Y_2(t) = \frac{N_1}{1+\exp\left(-a_1(t-\tau_1)+b_1\right)} + \frac{N_2}{1+\exp\left(-a_2(t-\tau_2)+b_2\right)}, \tag{6.3}$$

where $Y(t)$ number of total users, including generation 1 and generation 2, at the time t; $Y_1(t)$, number of users of generation 1 at the time t; $Y_2(t)$, number of users of generation 2 at the time t; N_1, constant potential of generation 1; N_2:,constant potential of generation 2; a_1, imitation coefficient of generation 1; a_2, imitation coefficient of generation 2; b_1, coefficient denoting the initial status of generation 1; b_2, coefficient denoting the initial status of generation 2; τ_1, time of introduction of generation 1; τ_2, time of introduction of generation 2.

6.2.1.3 Logistic Growth within a Dynamic Carrying Capacity Model

Not all technologies can be divided into clearly separate generations as the bi-logistic model assumes. Most of the time new functions are created gradually, added to the technology little by little. It is considered that the number of potential technological adopters increases whenever new functions are introduced. The model Sharif and Ramanathan [30] proposed has the upper bound increasing according to the logistic curve. In this paper it is called the logistic growth within a dynamic carrying capacity (LGDCC) model, which can be enumerated as (6.4):

$$\frac{\mathrm{d}Y(t)}{\mathrm{d}t} = \left(a\frac{Y(t)}{N(t)}\right)(N(t)-Y(t)). \tag{6.4}$$

In the LGDCC model, potential of the technology (carrying capacity) N is a function of time as depicted in (6.5).

$$\frac{\mathrm{d}N(t)}{\mathrm{d}t} = a_k\frac{N(t)}{N_k}(N_k - N(t)). \tag{6.5}$$

By synchronizing (6.4) and (6.5), LGDCC model can be developed as follows:

$$Y(t) = \frac{N_K}{1+\exp(-at+b)+\frac{a}{a-a_K}\exp(-a_K t+b_K)}, \tag{6.6}$$

where a_k and b_k, coefficients; and N_k, ultimate carrying capacity.

6.2.1.4 Choice-Based Substitution Diffusion Model

The utility that the *i*th potential customer (i.e. non-subscriber) would obtain by choosing not to subscribe or by subscribing to service at time t can be depicted as:

$$U_{ti}^{(0,k)} = V_t^{(0,k)} + \varepsilon_t^{(0,k)}, \quad k = 0, 1, 2, \tag{6.7}$$

where $k = 0$, 1, 2 indicate *non-subscription, generation 1 and 2 service subscription*, respectively.

In the superscript, the first term represents the subscription status of the individual just before the choice, and the second term represents the choice made at time t. Thus, the superscript (0,0) means that the ith *non-subscriber* remains a *non-subscriber*. The superscripts (0,1) and (0,2) mean that the ith *non-subscriber* chooses *generation 1 and 2 service* at time t, respectively. Similarly, the superscript (1,1) means that the ith *generation 1 subscriber* remains the same service and (1,2) that the consumer upgrades to generation 2 service.

In the equation, V and ε denote the deterministic term and the error term of the utility. Assume V is independent from the individual consumer and is dependent only on the attributes of each service (e.g. price, advertising, design, etc.), the error term, ε, is stochastic and captures both random taste variation across the population and model specification error. These assumptions make it possible to aggregate across individuals. In the second choice situation, the ith *generation 1 subscriber* at time t must decide whether to upgrade its service dependency. The utility that the ith *generation 1 subscriber* would obtain by choosing a specific alternative at time t can be depicted as follows:

$$U_{ti}^{(1,k)} = V_t^{(1,k)} + \varepsilon_t^{(1,k)},\ k = 1,2,3 \text{ (where three implies the coexistence of generation 1 and 2)} \tag{6.8}$$

Based on Jun and Park [16], the model for the ICT market in Japan is synthesized. In order to trace the diffusion and substitution dynamics for successive generations of products as well as for a single generation, the deterministic terms of the utility is specified as follows:

$$\text{When } t < \tau_2, V_t^{(0,0)} = C_{\tau 1}^{(0,0)} = C_{\tau 1}, V_t^{(0,1)} = Q_{r01} \times r_1, \tag{6.9}$$

where τ_2, time of introduction of generation 1; r_1, penetration rate in the preceding period; $C_{\tau_i}^{(0,0)}$, coefficient of $V_t^{(0,0)}$ condition at time τ_i $(i = 1,2)$; and $Q_{xjk}(x = r, t, f, p;\ j = 0, 1;$ and $k = 0, 1, 2, 3)$: coefficient of $V_t^{(j,k)}$ condition due to penetration (r), time trend (t), function (f) and price (p), respectively.

$$\begin{aligned} &\text{When } t \geqslant \tau_2, V_t^{(0,0)} = C_{\tau_2}^{(0,0)} = C_{\tau_2}, V_t^{(0,1)} = Q_{t01} \times (t - \tau_1 + 1), \\ V_t^{(1,1)} &= Q_{t11} \times (t - \tau_1 + 1), V_t^{(0,2)} = Q_{t02} \times (t - \tau_2 + 1) \\ &+ Q_{f02} \times f, V_t^{(1,2)} = Q_{12} \times (t - \tau_2 + 1) + Q_{f12} \times f + Q_{p12} \times p V_t^{(1,2)} \\ &= Q_{12} \times (t - \tau_2 + 1) + Q_{f12} \times f + Q_{p12} \times p. \end{aligned} \tag{6.10}$$

When a new product is introduced into the market, information about the product is uncertain and insufficient. However, as more information becomes available to consumers (***r*** increase), they can achieve higher levels of utility as product recognition increases. While some attributes, such as price (***p***) and function (***f***), are available, others are not easy to quantify or to observe even if they have a significant influence on the decision process of consumers. In such situations, the time variables (***t***) may explain the effect of factors such as design, sales promotion and fashion.

When the error terms ε is assumed to follow independent Gumbel distribution, the probability that a consumer i subscribes enjoy utility by generation 1 or 2 service at time t is:

$$P_t^{(0,k)} = \frac{\exp(V_t^{(0,k)})}{\exp(V_t^{(0,0)}) + \exp(V_t^{(0,1)}) + \exp(V_t^{(0,2)})}, \tag{6.11}$$

where the subscript i is omitted in the probability because the choice probability in (6.11) is the same for all individual under the assumption that the deterministic terms are independent from the individual. Similarly, the probability that *generation 1 subscriber* upgrades to *generation 2 service* at time t can be depicted by (6.12):

$$P_t^{(1,2)} = \frac{\exp(V_t^{(1,2)})}{\exp(V_t^{(1,1)}) + \exp(V_t^{(1,2)})}. \tag{6.12}$$

Define the total market potential of the choice based model at time t. N_t as (6.13) since it is possible that the market potential may be unchanged throughout the time period:

$$N_t = N_k, \tau_k \leqslant t < \tau_{k+1}. \tag{6.13}$$

The total number of subscribers both *generation 1 and 2 services* at time $t - 1$ is denoted as Y_{t-1}. Then the total number of *non-subscribers* at time t is $(N_t - Y_{t-1})$. Before the introduction of *generation 2 service* $(t < \tau_2)$, the expected net number of subscribers for *generation 1 service* at time t can be depicted:

$$\Delta Y_t^1 = (N_1 - Y_{t-1})P_t^{(0,1)} = (N_1 - Y_{t-1})\frac{\exp(V_t^{(0,1)})}{\exp(V_t^{(0,0)}) + \exp(V_t^{(0,1)})}. \tag{6.14}$$

After the introduction of *generation 2 service* $(t \geqslant \tau_2)$, the expected net number of subscribers at time t for each service can be defined as:

$$\Delta Y_t^1 = (N_2 - Y_{t-1})\frac{\exp(V_t^{(0,1)})}{\exp(V_t^{(0,0)}) + \exp(V_t^{(0,1)}) + \exp(V_t^{(0,2)})} - Y_{t-1}^1 \frac{\exp(V_t^{(1,2)})}{\exp(V_t^{(1,1)}) + \exp(V_t^{(1,2)})}, \tag{6.15}$$

$$\Delta Y_t^2 = (N_2 - Y_{t-1})\frac{\exp(V_t^{(0,2)})}{\exp(V_t^{(0,0)}) + \exp(V_t^{(0,1)}) + \exp(V_t^{(0,2)})} + Y_{t-1}^1 \frac{\exp(V_t^{(1,2)})}{\exp(V_t^{(1,1)}) + \exp(V_t^{(1,2)})}, \tag{6.16}$$

where $Y_t^k = Y_{t-1}^k + \Delta Y_t^k$, $k = 1, 2$ and $Y_t = Y_t^1 + Y_t^2$.

Equation (6.15) denotes the increase of subscribers who were previous *non-subscribers*, while (6.16) denotes the upgraders who switch from *generation 1 service to generation 2 service*.

6.2.2 Data Construction

Data for subscriber base of mobile phone service in Japan, including NonIP and IP mobile phone, is published by Telecommunications Carriers Association (TCA) Japan [31]. Data for subscriber base of the internet connection service, including dial-up, ISDN, DSL and optical fiber, is published by the Ministry of Public Management, Home Affairs, Posts and Telecommunications (MPHPT) Japan [32]. Further data about ICT in Japan is retrieved from IT White Paper published by Japan Information Processing Development Corporation and from annual survey result of ICT usage style [33].

Overall data worldwide, including population and other indicators such as GDP per capita, is retrieved from World Development Indicators published by World Bank [5]. Digital Access Index of each country is published by ITU [4].

The empirical analysis, including multi-factor analysis, nonlinear regression and correlation analysis is conducted with the statistical software, SPSS.

6.3 Diffusion and Substitution Process of ICT in Japan: Empirical Analysis with Diffusion and Substitution Models

6.3.1 Telephony Market in Japan

6.3.1.1 Fixed Phone Line Diffusion Trend in Japan

In order to analyze telephony market in Japan, first, by using simple logistic model and choice-based substitution diffusion model as synthesized in Sect. 6.2, fixed phone line diffusion trend is analyzed.

(1) *Simple logistic model*
Similar to the situation in other countries, fixed line phone service had been continuing its steady progress in Japan until 1996. However, it turned down since then with its peak in 1996, claiming that it is no longer an isolated ICT industry that could spread out as it had. This is demonstrated by the simple logistic model using the TCA's empirical data. As summarized in Table 6.1 and illustrated in Fig. 6.2, the potential (carrying capacity) of fixed phone line subscription is anticipated to 61.8 million, equivalent level of the diffusion in 1996.

(2) *Choice-based substitution diffusion model*
The possible reason for such dramatic inflection point followed by decline in fixed line phone service can be attributed to the emergence of the mobile phone service. As mentioned in Sect. 6.1, mobile phone and fixed phone may complement and substitute each other. In case of the complement, fixed phone can survive, while the substitution leads it to obsolescent. In order to elucidate this dynamism, the markets of both services are analyzed by using the choice-based substitution diffusion model.

Table 6.1 Estimation results of the fixed phone line subscription diffusion in Japan by simple logistic model (1953–2002)

Estimated model $Y(t) = \frac{N}{1+\exp(-at+b)}$			
Parameter	Estimate	t-value	adj.R^2
a	1.52×10^{-1}	4.26	0.991
b	3.70×10^{0}	3.56	
N	6.18×10^{3}	6.45	

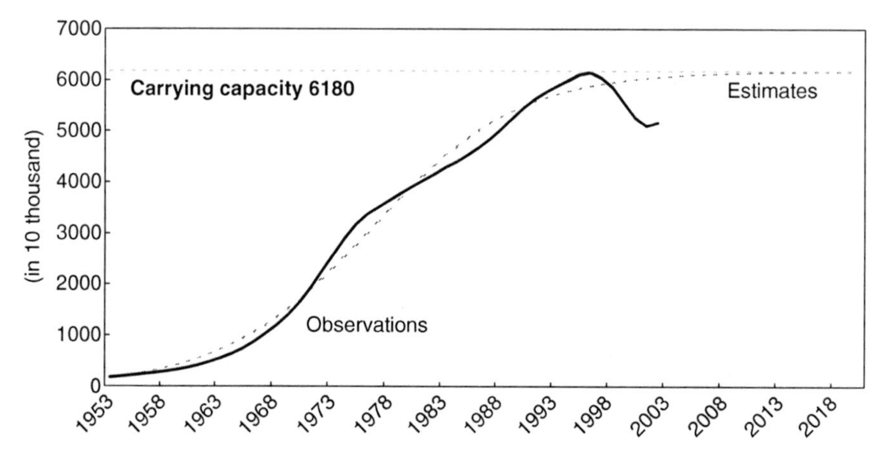

Fig. 6.2 Observations[a] and estimates[b] of the fixed phone line subscription diffusion in Japan by simple logistic model
[a] 1953–2002
[b] 1953–2020
[c] Original statistics are based on monthly reports by TCA, Japan [31]

While the difficulty in taking both services into a single model lies in the double counting of both fixed and mobile phones subscribers, and it is generally difficult to identify such double subscribers, a solution was obtained by depending on the survey conducted by MPHPT [32].

Similar to the model explained in Sect. 6.2 and the model to be used in Sects. 6.3.2 and 6.3.3 for mobile phone and the internet access analyses, Y_t^1 denotes the first generation technology (in this case, fixed phone line) and Y_t^2 denotes the second generation technology (similarly, mobile phone). τ_1 and τ_2 indicates the time of introduction of fixed phone and mobile phone, respectively. Impulses for substitution diffusion are penetration of fixed phone line (r_{01}), time trend (t_{01}) and price (p_{jk} where $j = 0, 1$ and $k = 2, 3$) are taken. Q_{p13} indicates the coefficient of fixed phone subscriber's adoption of both mobile phone and fixed phone due to price (p) reason. It is assumed that non-subscribers first sign up to only one service, and fixed-line-only user may switch to mobile phone service but not vice versa. The concept of this switching is illustrated in Fig. 6.3.

The estimated result is summarized in Table 6.2 and illustrated in Fig. 6.4. Looking at Table 6.2 we note that while fixed line phone penetrates into non-subscriber

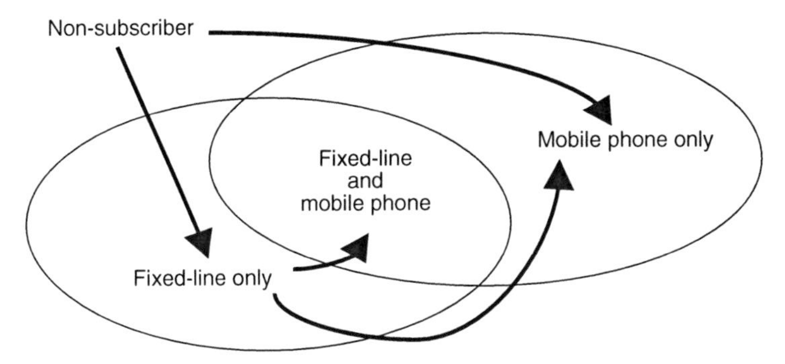

Fig. 6.3 Switching among different groups in telephony market

Table 6.2 Estimation results of the substitution trajectory of mobile subscribers for fixed line in Japan by choice-based substitution diffusion model

Parameter[a]	Estimate	t-value	adj.R^2
C_{τ_1}	1.22×10^2	1.86	0.994
C_{τ_2}	6.50×10^{-1}	1.97	
N_1	5.74×10^3	5.76	
N_2	10.35×10^3	3.22	
Q_{r01}	1.67×10^0	2.67	
Q_{t01}	-1.00×10^1	3.12	
Q_{p02}	-4.72×10^{-1}	2.43	
Q_{p12}	-1.06×10^0	1.98	
Q_{p13}	-3.02×10^0	2.43	

[a] $\boldsymbol{C\tau_1}$, $\boldsymbol{C\tau_2}$, coefficients of the initial utility before and after the introduction of mobile phone; $\boldsymbol{N_1}$, $\boldsymbol{N_2}$, potential market of fixed phone and mobile phone; $\boldsymbol{Q_{r01}}$, coefficient of non-subscriber chooses fixed phone due to its market penetration; $\boldsymbol{Q_{t01}}$, fixed phone subscriber chooses mobile phone due to time trend; and $\boldsymbol{Q_{p02}}$, $\boldsymbol{Q_{p12}}$, $\boldsymbol{Q_{p13}}$, coefficient of non-subscriber and fixed phone subscriber choose mobile phone or both fixed and mobile phones, respectively due to price

($Q_{\mathrm{r}01} = 1.67$), its attractiveness decreases as time passes ($Q_{\mathrm{t}01} = -1.00$). Price decrease induces mobile phone subscriber from non-subscriber ($Q_{\mathrm{p}02} = -0.47$), shifting from fixed phone ($Q_{\mathrm{p}12} = -1.06$ and $Q_{\mathrm{p}13} = -3.02$). Figure 6.4 demonstrates clearly that mobile phone emerged in Japan's ICT market in 1991 and exhibited a dramatic deployment exceeding the diffusion level of fixed phone in 1998 and approaching to its carrying capacity 103.5 million (N_2), 1.7 times higher than the capacity of fixed phone. While this demonstrates that mobile phone substitutes for fixed phone, Table 6.2 suggests that the potential of fixed phone (N_1) is 57.4 million and 7% lower than the estimate of the simple logistic model. This balance suggests the co-existence of both fixed and mobile phones in the initial stage of the emergence of mobile phone due to the price decrease of telephony service by the emergence of mobile service. A coefficient $Q_{p13}(-3.02)$ demonstrates this dynamism. Thus, we could conclude that mobile phone is substituting for fixed phone with certain complement induced by price decrease.

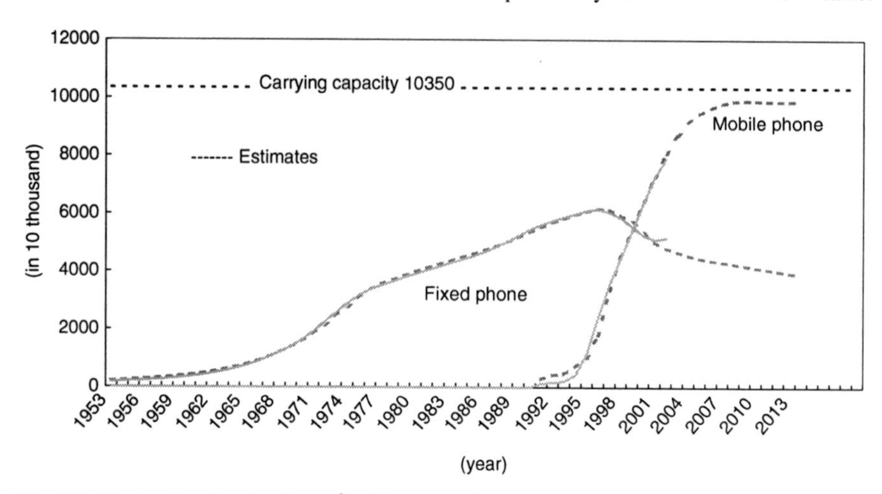

Fig. 6.4 Observations[a] and estimates[b] of the substitution trajectory of mobile subscribers for fixed line in Japan by choice-based substitution diffusion model
[a] 1953–2002
[b] 1953–2015
[c] Original statistics are based on monthly reports by MPHPT, Japan [32]

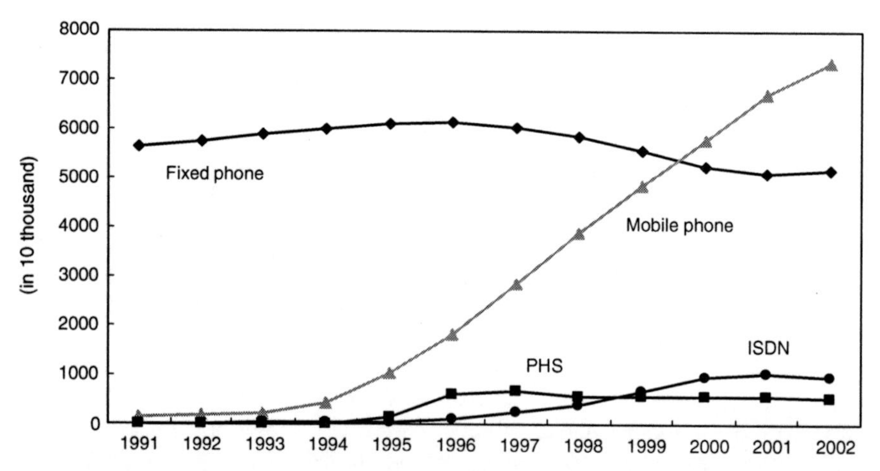

Fig. 6.5 Trends in number of "fixed mainline," "mobile phone," "PHS" and "ISDN" Subscribers in Japan (1991–2002)
Source: MPHPT, Japan [32]

6.3.1.2 From Fixed to Mobile Telephony, and From Voice to Data

The substitutability of mobile phone to replace fixed phone can be attributed largely to (a) its functionality of not only voice but also data, (b) belonging to individual from sharing by family, and (c) mobility, particularly, no need to change telephone numbers whenever customers move.

With such understanding, trends in number of telephony service in Japan by type are illustrated in Fig. 6.5 by compiling the MPHPT's data.

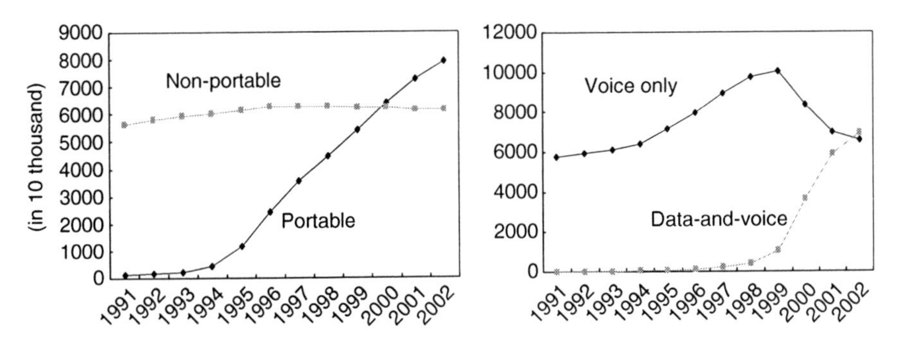

Fig. 6.6 Trends in number of portable and non-portable telephony subscription and number of voice-only and data-and-voice telephony subscription in Japan (1991–2002)
Source: Compiled base on data from MPHPT, Japan [32]

The trends in portable and non-portable telephony subscription as well as voice-only and data and voice telephony subscription can be compared as illustrated in Fig. 6.6. Portable category contains mobile phone and PHS services while non-portable represents fixed phone and ISDN. Voice-only category contains fixed mainline while data-and-voice consists of ISDN, mobile phone and PHS. All demonstrates the significance of a mobile phone market in Japan.

6.3.2 Mobile Telephony Market in Japan

6.3.2.1 Restructured Telecom Industry

When NTT DoCoMo first introduced i-mode service in February 1999, only 5,000 out of 40.5 million mobile subscribers registered for the service. Following NTT DoCoMo, au and J-PHONE started mobile internet service "ezweb" and "J-Sky" in April and December 1999, respectively. This resulted in price reduction and new service leading to mobile internet subscriber base (internet protocol (IP) mobile subscribers) surging to 67.8 million, occupying 84.98% of mobile subscriber base as demonstrated in (Figs. 6.7 and 6.8).

The success of mobile internet has provided an opportunity of shifting the mobile phone usage from voice calls to non-voice communications as demonstrated in Table 6.3.

6.3.2.2 Mobile Phone Diffusion Dynamism in Japan

In order to analyze the mobile phone diffusion dynamism in Japan. Simple logistic model, bi-logistic model, choice-based substitution diffusion model, and logistic growth within a dynamic carrying capacity model are used.

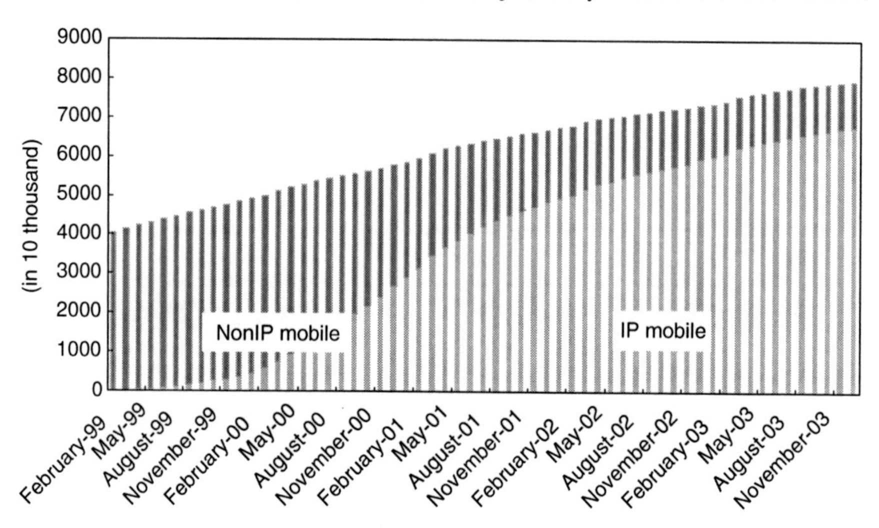

Fig. 6.7 Trends in number of NonIP and IP mobile subscribers in Japan (Feb. 1999–Dec. 2003)
Source: Telecommunications Carriers Association, Japan [31]

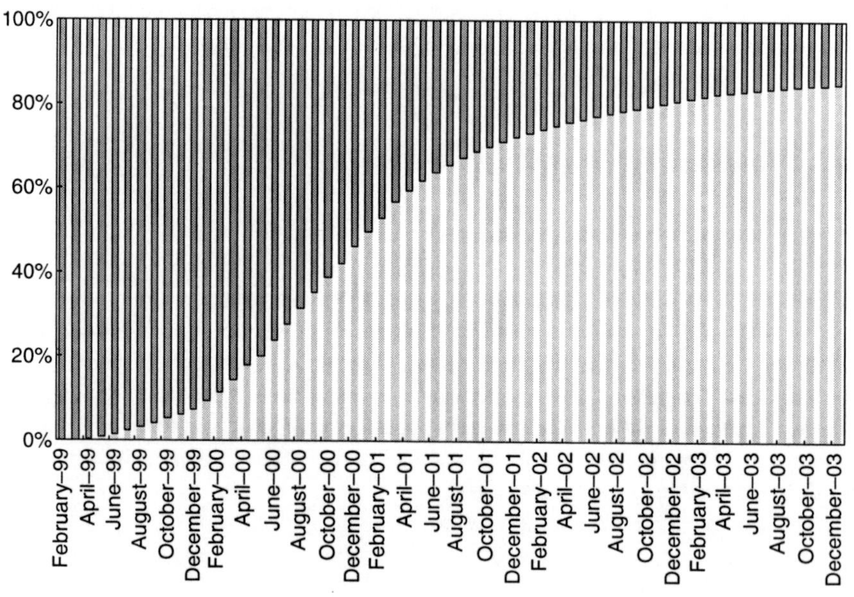

Fig. 6.8 Share of NonIP and IP mobile subscribers in Japan (Feb. 1999–Dec. 2003)
Source: Telecommunications Carriers Association, Japan [31]

(1) *Simple logistic model*

The result of taking the whole mobile market to fit the simple logistic model is summarized in Table 6.4 and also illustrated in Fig. 6.9, with observations (Jan. 1996–Dec. 2003) and estimates (Jan. 1996–Jan. 2008).

Table 6.3 Trend in new services and functions introduced by mobile phone carriers (1999–2003)

Time	New service/function	Carrier'
Feb 1999	i mode	NTTDoCoMo
Apr 1999	Ezweb	Au
Dec 1999	J-sky	J-PHONE*
Jan 2000	DocoNavi	NTTDoCoMo
May 2000	J-Navi	J-PHONE*
Nov 2000	J-pic-mail	J-PHONE*
Jan 2001	i-appli (Java)	NTTDoCoMo
Jun 2001	JgalvaAppli(JAva)	J-PHONE*
Jul 2002	ezplus (Java)	Au
Nov 2001	i-motion (video clip)	NTTDoCoMo
Dec 2001	ezmovie (video clip)	Au
Dec 2001	eznavi (navigation)	Au
May 2002	i-shot (camera)	NTTDoCoMo
May 2002	au-shot (camera)	Au
Dec 2002	J-movie (video clip)	J-PHONE*
Dec 2003	Mobile TV	Vodafone E

Sources: IT White Book [33]

Table 6.4 Estimation results of the mobile subscriptions in Japan by simple logistic model (Jan. 1996–Dec. 2003)

Estimated model $Y(t) = \frac{N}{1+\exp(-at+b)}$			
Parameter	Estimate	t-value	adj.R^2
a	4.44×10^{-2}	5.22	0.996
b	5.63	7.31	
N	8.47×10^{-3}	10.24	

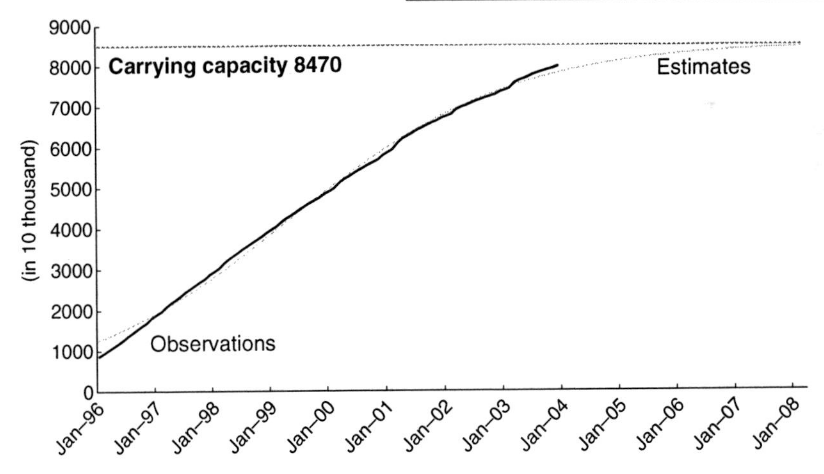

Fig. 6.9 Observations[a] and estimates[b] of mobile subscribers in Japan by simple logistic model: in 10,000
[a] Jan. 1996–Dec. 2003
[b] Jan. 1996–Jan 2008
[c] Original statistics are based on monthly reports by TCA, Japan [31]
Source: Telecommunications Carriers Association, Japan [31]

Table 6.5 Estimation results of the mobile subscriptions in Japan by bi-logistic model (Jan. 1996–Dec. 2003)

Estimated model $Y(t) = Y_1(t) + Y_2(t) = \frac{N_1}{1+\exp(-a_1(t-\tau_1)+b_1)} + \frac{N_2}{1+\exp(-a_2(t-\tau_2)+b_2)}$

Parameter	Estimate	t-value	adj.R^2
a_1	7.66×10^{-2}	8.27	0.998
b_1	8.00	12.31	
a_2	8.62×10^{-2}	5.56	
b_2	2.68	10.15	
N_1	4.91×10^{-3}	15.21	
N_2	3.23×10^{3}	13.38	

By means of the interpretation of the result, the recent phenomenon of slowing-down growth can be explained as an outcome of saturated market. The time required to reach half of the potential is b/a, which equals 126.8 (months), and this indicates that Japan took only about 10 years (since 1988) to reach the half point. While the similar analysis is conducted with the data of high-income countries, the time required to reach 50% of potential is about 13 years. Compared with the average of high-income countries, Japan is a relatively active mobile market.

(2) *Bi-logistic model*

Although taking the mobile market as a whole helps to understand the overall vision, it is necessary to separate the mobile market into two stages: before and after the mobile internet was introduced. With the bi-logistic model, it is attempted to capture the two impulses inside the market. Let N_1 and N_2 be the potential of NonIP mobile and IP mobile diffusion, respectively. The result of bi-logistic model analysis is summarized in Table 6.5 and also illustrated in Fig. 6.10, with the observations of the overall market size (Jan. 1996–Dec. 2003) along with the estimates (Jan. 1996–Jan. 2008) representing for the whole market, NonIP impulse followed by the IP impulse.

Judging from the parameters obtained from the model, the total diffusion process is divided into two impulses. The first impulse came from the mobile service only capable of voice calls transmission. The potential is approximately 49.1 million, which means without the mobile internet service, it was possible that only half of the population would have been using mobile phone. However, earning growth due to mobile service capable of non-voice communication, the potential rises up by 32.3 million, pulling high the upper limit of mobile phone diffusion to 81.4 million.

(3) *Choice-based substitution diffusion model*

Base on the result of bi-logistic model analysis, it is plausible that the dramatic increase in IP mobile subscriber base does not come from the second impulse only. There must have been internal mechanism pushing IP mobile service to spread out so fast. Therefore, it is considered that those previously existing users switched from NonIP to IP mobile service after the introduction of IP mobile. Thus, a diffusion trajectory by means of choice-based substitution diffusion model is used to analyze this internal change after IP mobile phone was introduced.

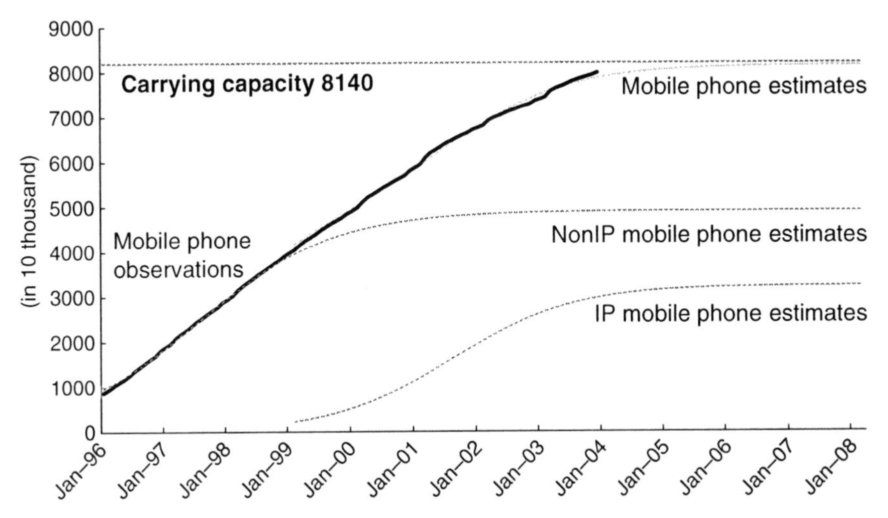

Fig. 6.10 Observations[a] and estimates[b] of mobile subscribers in Japan by bi-logistic model
[a] Jan. 1996–Dec. 2003
[b] Jan. 1996–Jan. 2008
[c] Original statistics are based on monthly reports by TCA, Japan [31]

As explained in Sect. 6.2, based on the existing studies, factors such as penetration (r), time trend (t), price (p) and function (f) are taken into the model. The choices made by consumers consist of the following possibilities: (a) before IP mobile is introduced, non-subscribers sign up to mobile service or not; (b) after IP mobile was introduced, non-subscribers sign up to mobile service or not; (c) they can choose NonIP or IP service if they sign up to mobile service; and (d) existing subscribers may switch from NonIP to IP service. Thus, IP mobile gain more population from new subscribers as well as existing subscribers.

The result of choice-based diffusion model analysis is summarized in Table 6.6 and also illustrated in Fig. 6.11, with the observations of the overall market size (Jan. 1996–Dec. 2003) along with the estimates (Jan. 1996–Jan. 2007) representing NonIP mobile and IP mobile service subscriber bases.

Looking at Table 6.6 we note that while NonIP mobile phone penetrates into non-subscriber ($Q_{r01} = 6.17$), its attractiveness decreases as time passes ($Q_{t01} = -1.77$). Contrary to NonIP, IP mobile gains popularity and attracts customers from non-subscriber and IP mobile subscriber as time passes ($Q_{t02} = 0.12$, $Q_{12} = 0.02$). In addition, as function improves, subscriber rushes into IP mobile ($Q_{f02} = 0.005$, $Q_{f12} = 0.005$). Consequently, price decrease induces a shift from NonIP to IP mobile subscriber ($Q_{p12} = -8.00$). Figure 6.16 demonstrates clearly that IP mobile phone emerged in Japan's ICT market in 1999 and substitutes NonIP dramatically leading to carrying capacity of mobile phone 82.3 million (N_2). This estimate is almost the same as the estimate by bi-logistic model (81.4 million). However, the estimation by choice-based substitution diffusion model expects stronger substitution of IP for NonIP mobile phone subscriber.

Table 6.6 Estimation results of the mobile subscriptions in Japan by choice-based substitution diffusion model (Jan. 1996–Dec. 2003)

Parameter[a]	Estimate	t-value	adj.R^2
C_{τ_1}	7.65×10^1	5.20	0.997
$C_{\tau 2}$	7.43×10^1	3.51	
N_1	5.57×10^3	1.85	
N_2	8.23×10^3	6.87	
Q_{r01}	6.17×10^0	2.35	
Q_{t01}	-1.77×10^{-1}	7.52	
Q_{t02}	1.16×10^{-1}	3.45	
Q_{t11}	1.60×10^{-2}	2.65	
Q_{t12}	2.00×10^{-2}	3.21	
Q_{f02}	4.61×10^{-3}	1.86	
Q_{f12}	5.00×10^{-3}	3.12	
Q_{p12}	-8.00×10^0	5.15	

[a] $C\tau_1$, $C\tau_2$, coefficients of the initial utility before and after the introduction of mobile phone; N_1, N_2, potential market of NonIP and IP mobile phone; Q_{r01}, coefficient of non-subscriber chooses NonIP mobile phone due to its market penetration; Q_{t01}, Q_{t02}, non-subscriber chooses NonIP and IP mobile phone, respectively due to time trend; Q_{t11}, Q_{t12}, similarly, NonIP subscriber chooses NonIP and IP mobile phone, respectively due to time trend; Q_{f02}, Q_{f12}, coefficient of non-subscriber chooses mobile phone, NonIP subscriber chooses IP mobile, respectively due to function; and Q_{p12}, NonIP subscriber chooses IP mobile due to price

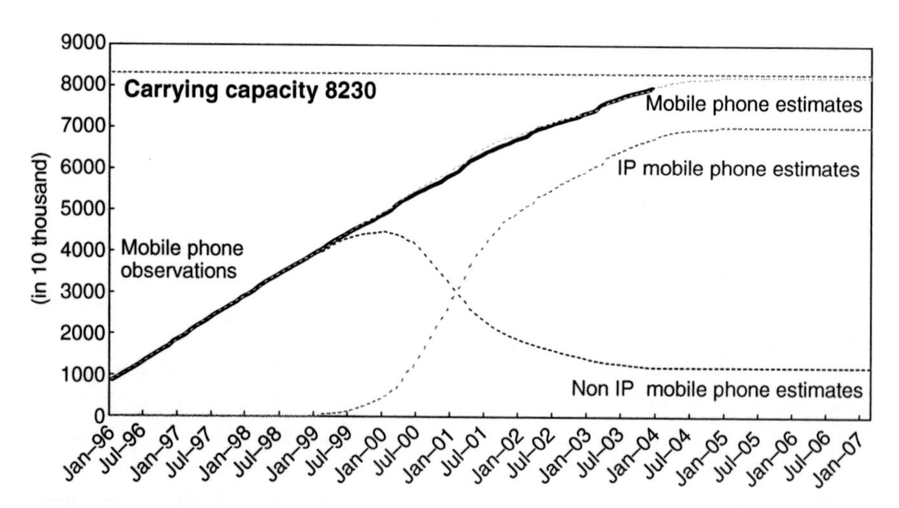

Fig. 6.11 Observations[a] and estimates[b] of NonIP and IP mobile subscribers in Japan by choice-based substitution diffusion model

[a] Jan. 1996–Dec. 2003

[b] Jan. 1996–Jan. 2007

[c] Original statistics are based on monthly reports by TCA, Japan [31]

Next, the positive M shows that in the first stage, as the penetration rate increases, consumers' utility of adopting mobile service rises; this can also be regarded as the so-called "imitation" in Bass model. However, the negative Q_{01} identifies that, once the second stage starts, non-subscribers' utility of adopting NonIP mobile service decreases as time goes by because they have one more choice, IP mobile service. NonIP mobile offers less and less utility to users; NonIP mobile service becomes obsolete after mobile internet enters the market. On the contrary, the positive Q_{02} points out that, for non-subscribers, IP mobile service seems to be the main mobile service as time passes; therefore, the later they decide to sign up to mobile service, the more possibly they would sign up to IP mobile phone directly.

(4) *Logistic growth within a dynamic carrying capacity*

With the result of choice-based substitution diffusion model, it is appropriate to assume that in the mobile market in Japan the potential of market grows with the evolving technology. Although in the empirical study of choice-based substitution diffusion model the diffusion process is only divided into two phases, technology innovation and development can also be considered as a continuous process that is difficult to draw a line in between two generations. Whenever the performance is improved or the product is upgraded, the potential is lifted up in some degree. Therefore, the logistic growth within a dynamic carrying capacity is introduced to deal with such continuous innovation process. Taking the mobile phone market as a whole to observe its growth since December 1988, the diffusion trajectory is analyzed and the result is summarized in Table 6.7 and also illustrated in Fig. 6.12.

The final potential of mobile phone market estimated by the logistic growth within a dynamic carrying capacity reaches about 93.8 million, higher than potentials previously estimated by other models. Moreover, while the simple logistic model and the bi-logistic model seem underestimate the growth after year 2004, the logistic growth within a dynamic carrying capacity seems to reflect the growing feature of market potential due to expanded functions, reduced prices, increased population or other non-specific factors.

As long as the market mechanism and mobile technology innovation work smoothly hereafter as expected, the current slowing-down market growth will still climb up to fulfill the final potential.

Table 6.7 Estimation results of the mobile subscriptions in Japan by logistic growth within a dynamic carrying capacity (Jan. 1996–Dec. 2003)

Estimated model $Y(t) = \frac{N_N}{1+\exp(-a't+b')+\frac{a'}{a'-a'_N}\exp(-a'_Nt+b_N)}$

Parameter	Estimate	t-value	adj.R^2
a'	1.48×10^{-1}	8.94	0.998
b'	1.44×10^{2}	9.85	
a_K	3.43×10^{-2}	7.63	
b_K	4.21×10^{0}	8.65	
N_N	9.38×10^{3}	10.89	

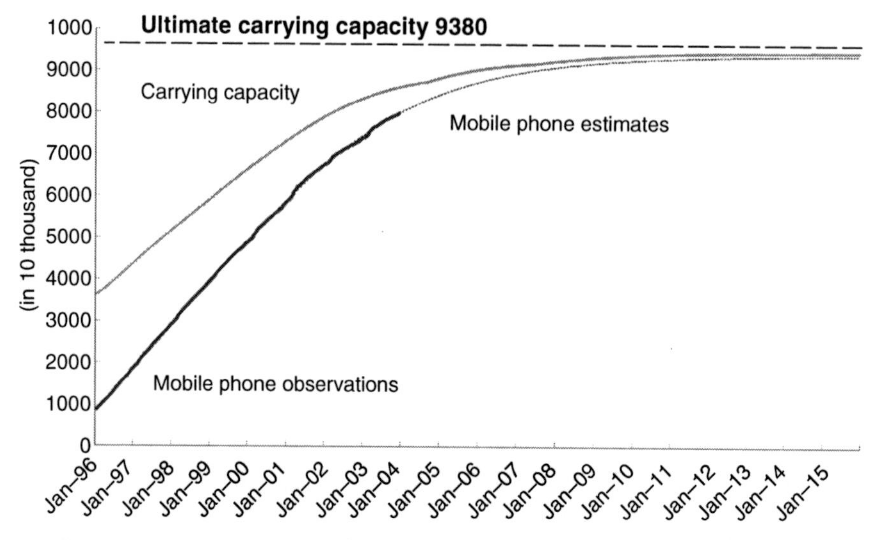

Fig. 6.12 Observations[a] and estimates[b] of mobile subscribers in Japan by logistic growth within a dynamic carrying capacity

[a] Jan. 1996–Dec. 2003

[b] Jan. 1996–Dec. 2015

[c] Original statistics are based on monthly reports by TCA, Japan [31]

6.3.3 Internet Access in Market Japan

6.3.3.1 Diffusion of the Internet Access in Japan

Figure 6.13 demonstrates trend in number of the internet access subscribers in Japan by classifying PC-terminal internet access, dial-up access and broadband access. Figure 6.14 demonstrates service of mobile internet subscribers in Japan used by the customers.

In order to explore how competition and price reduction affected the internet market in Japan in the past few years, numerical analysis are conducted by applying single logistic model, bi-logistic model and choice-based substitution diffusion model.

(1) *Simple logistic model*

The result of fitting the empirical data of internet subscription in Japan with simple logistic model shows its diffusion potential is approximately 37 million as summarized in Table 6.8 and demonstrated in Fig. 6.15.

Compared with the potential of mobile phone market potential, which is about 85 million, the potential of internet subscription seems to be much smaller. This can be attributed to the fact that mobile phone service, both the handset and the phone number, is a kind of telecom service for each individual. People rarely share their mobile phone with any one else; however, in the case of internet, especially all internet accesses other than wireless internet, subscribers usually install the devices

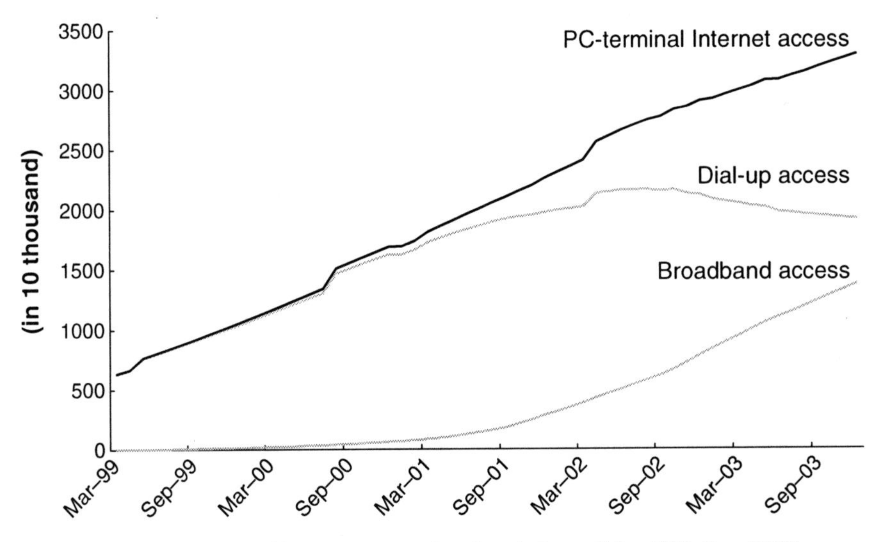

Fig. 6.13 Trend in number of internet access subscribers in Japan (Mar. 1999–Dec. 2003)
Source: MPHPT, Japan [32]

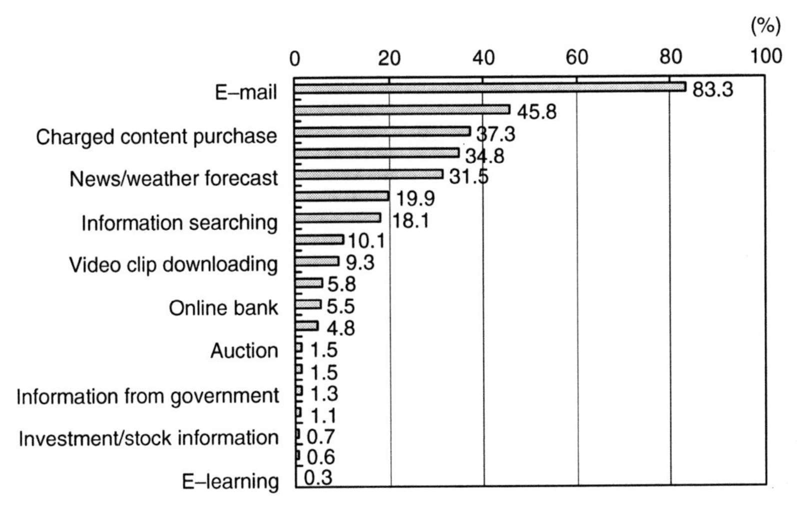

Fig. 6.14 Service used by mobile internet subscribers in Japan
Source: Telecommunications Usage Trend Survey 2003 [32]

(cable, telephone line, personal computer, etc.) at home and share with their family members. It is quite understandable that internet subscription potential is lower than that of mobile phone. However, the diffusion speed b/a of internet access does not reveal significant difference from that of the mobile phone.

Table 6.8 Estimation results of internet access subscribers in Japan by simple logistic model (Oct. 1999–Dec. 2003)

Estimated model $Y(t) = \frac{N}{1+\exp(-at+b)}$			
Parameter	Estimate	t-value	adj.R^2
a	6.15×10^{-2}	6.58	0.995
b	7.61×10^{0}	7.46	
N	3.71×10^{3}	9.44	

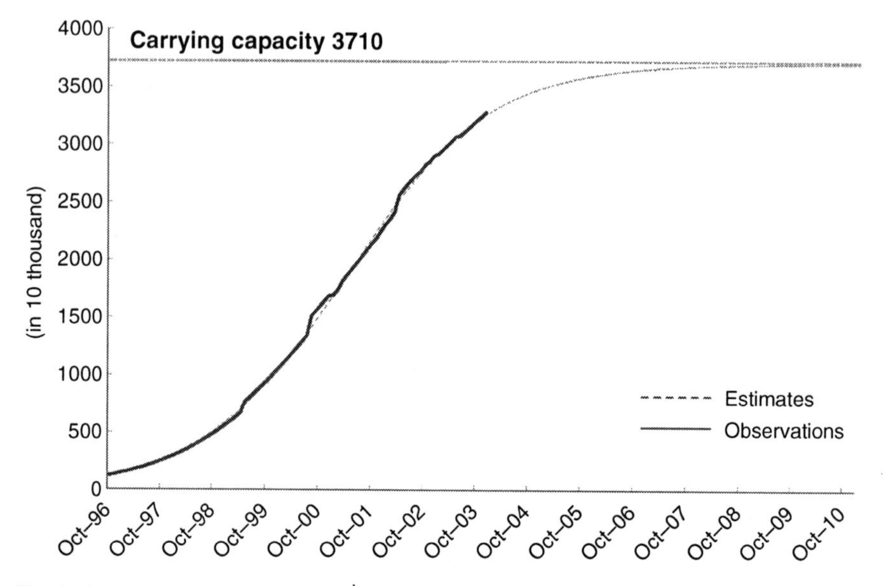

Fig. 6.15 Observations[a] and estimates[b] of internet access subscribers in Japan by simple logistic model

[a] Oct. 1999–Dec. 2003

[b] Oct. 1999–Dec. 2010

[c] Original statistics are based on monthly reports by MPHPT, Japan [32]

(2) *Bi-logistic model*

While the simple logistic model analysis gives a general outline about this market in Japan, aiming at exploring its internal change, the bi-logistic model is applied to analyze the two impulses of internet accesses, namely "dial-up" and "broadband." The start of dial up, τ_1, is identified as January 1991 while the start of broadband service, τ_2, is identified as March 1999. The result and estimated parameters are summarized in Table 6.9 and also illustrated in Fig. 6.16.

Table 6.9 and Fig. 6.16 demonstrate that the carrying capacity of Japan's internet subscribers is estimated 35.6 million composed of 23.5 million by dial-up access and 12.1 million by broadband access.

Table 6.9 Estimation results of internet access subscribers in Japan by bi-logistic model (Oct. 1999–Dec. 2003)

Estimated model

$$Y(t) = Y_1(t) + Y_2(t) = \frac{N_1}{1+\exp(-a_1(t-\tau_1)+b_1)} + \frac{N_2}{1+\exp(-a_2(t-\tau_2)+b_2)}$$

where Y_1, dial-up subscription; Y_2, broadband subscription

Parameter	Estimate	t-value	adj.R^2
a_1	6.28×10^{-2}	2.98	0.997
b_1	7.72	6.45	
a_2	9.42×10^{-2}	7.86	
b_2	3.69	7.47	
N_1	2.35×10^3	7.26	
N_2	1.21×10^3	4.87	

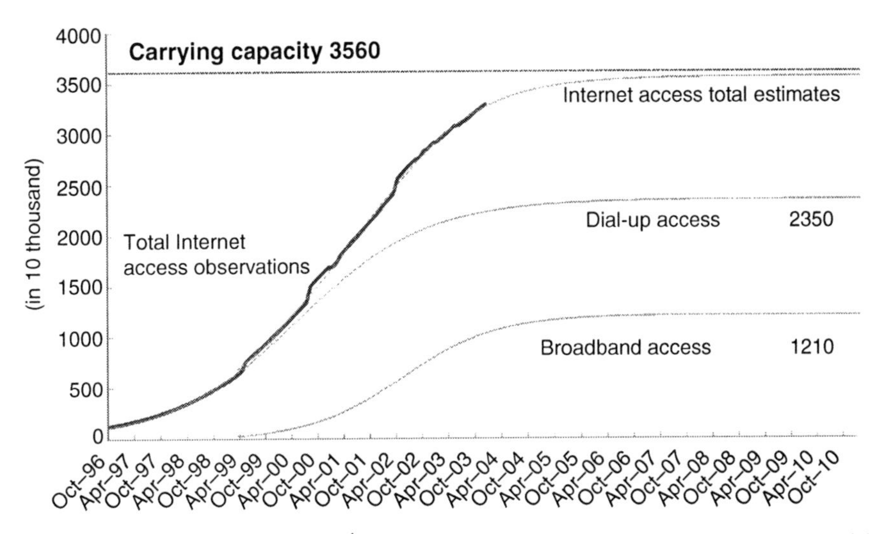

Fig. 6.16 Observations[a] and estimates[b] of internet access subscribers in Japan by bi-logistic model

[a] Oct. 1999–Dec. 2003

[b] Oct. 1999–Dec. 2010

[c] Original statistics are based on monthly reports by MPHPT, Japan [32]

(3) *Choice-based substitution diffusion model*

While contributing to the whole subscription potential, broadband service also attracts subscribers from the existing dial-up service user base, in order to decompose this mechanism of the dynamic change of both dial-up and broadband service diffusion. Choice-based substitution diffusion model is applied. Let the start of dial up, τ_1, as Jan. 1991 and the start of broadband service, τ_2, as Mar. 1999. As the model implemented in the previous sections, suppose generation 1 represents the dial-up access and generation 2 the broadband. Similar to preceding analyses, impulses of penetration (r), time trend (t), and price (p) are examined.

Table 6.10 Estimation results of the substitutive trajectory among internet, dial-up, and broadband access subscribers in Japan by choice-based substitution diffusion model (Oct. 1999–Dec. 2003)

Parameter[a]	Estimate	t-value	adj.R^2
$C_{\tau 1}$	2.30×10^2	2.02	0.997
C_{τ_2}	4.01×10^2	1.87	
N_1	7.75×10^2	2.14	
N_2	3.96×10^3	7.15	
Q_{r01}	9.71×10^{-2}	1.98	
Q_{t01}	2.71×10^{-2}	3.67	
Q_{t02}	1.90×10^{-1}	2.12	
Q_{t11}	5.00×10^{-2}	2.65	
Q_{t12}	8.51×10^{-2}	2.64	
Q_{p02}	-3.02×10^0	1.85	
Q_{p12}	7.21×10^{-1}	2.75	

[a] $C\tau_1, C\tau_2$, coefficients of the initial utility before and after the introduction of broad band; N_1, N_2, potential market of dial-up and broadband; Q_{r01}, coefficient of non-subscriber chooses dial-up due to its market penetration; Q_{t01}, Q_{t02}, coefficient of non-subscriber chooses dial-up and broadband, respectively, due to time trend; Q_{t11}, Q_{t12}, coefficient of dial-up subscriber chooses dial-up and broadband, respectively due to time trend; and Q_{p02}, Q_{p12}, coefficient of non-subscriber chooses broadband and dial-up subscriber chooses broad, respectively due to price

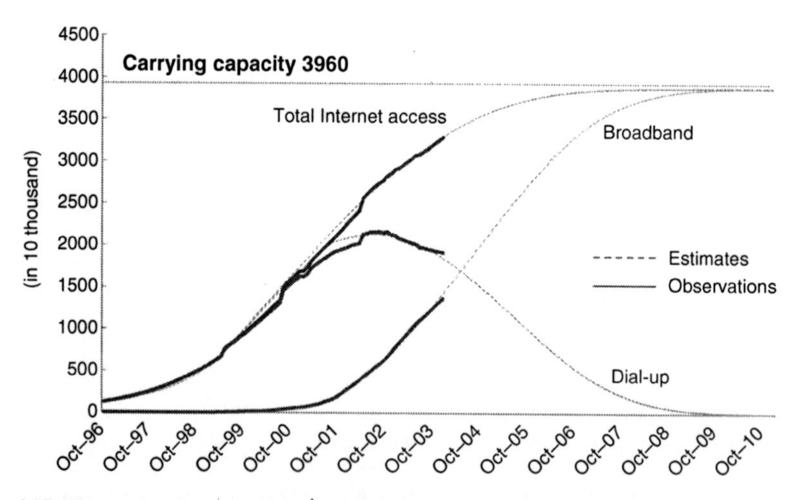

Fig. 6.17 Observations[a] and estimates[b] of internet, dial-up, and broadband access subscribers in Japan by choice-based substitution diffusion model

[a] Oct. 1999–Dec. 2003

[b] Oct. 1999–Dec. 2010

[c] Original statistics are based on monthly reports by MPHPT, Japan [32]

The estimated result is summarized in Table 6.10 and illustrated in Fig. 6.17, depicting the curve of dial-up diffusion and broadband diffusion estimates (Oct. 1999–Dec. 2010) as well as empirical observations (Oct. 1999–Dec. 2003).

Looking at Table 6.10 we note that dial-up penetrates into non-subscriber ($Q_{r01} = 0.10$), its attractiveness shifts to broadband as time passes ($Q_{t01} = 0.03$, $Q_{t02} = 0.19$, $Q_{t11} = 0.05$, $Q_{t12} = 0.09$). Price decrease induces non-subscriber to broadband ($Q_{p02} = -3.02$) while dial-up subscriber shifts to broadband despite its price increase ($Q_{p12} = 0.72$) as dial-up subscriber has realized the attractiveness of the high quality of broadband.

Before broadband service was offered, the market potential was about 7.8 million ($N_1 = 7.75 \times 10^6$), with the improved capacity of broadband, more consumers are likely to adopt internet access than before enhancing potential market by broadband 39.6 million ($N_2 = 3.96 \times 10^7$). In the first stage when only dial-up service existed, positive Q_{r01} shows that consumers' utility increased as the subscribers became more. In the second stage with both services, consumers' utility of using broadband service increases with time faster than that of using dial-up service since $Q_{t01} < Q_{t02}$. Furthermore, consumers' utility of switching from dial-up to broadband service increases with time faster than that of keeping using dial-up service since $Q_{t11} < Q_{t12}$. The price factor affects the non-subscribers more (when they decide to sign up to broadband service) than the existing dial-up users (when they decide to switch to broadband service) since $|Q_{p12}| < |Q_{p02}|$. All demonstrates the significance of the advancement of broadband in leveraging internet access.

6.3.3.2 Comparison Between Broadband Diffusion in Japan and Other Countries

With the understanding of the significance of the advancement of broadband in leveraging internet access, Fig. 6.18 analyzes the correlation between GNI (Gross National Income) per capita and broadband penetration rate in 30 countries the world which demonstrates that broadband penetration rate increases logistically as GNI per capita increases.

With South Korea as the outstanding outlier, this correlation can be described as an exponential function which indicates that Japan actually lies over the average level. Japan, one of the leading well-developed economies, has attained the strength for broadband diffusion, such as high GNI, relatively low price and high access speed. Most important feature is that it has been the most successful country in mobile telephony diffusion, just as South Korea in broadband diffusion. Its success in mobile telephony shows that Japan is a market capable of ICT adoption. After "Yahoo BB!" entered the market in September 2001, the increase of subscriber number surged up over 200,000 per month immediately and the broadband population increased by over 30,000 each month afterwards. Currently Japan still lags behind South Korea due to the late start in constructing competitive circumstances. However, with steady growth in competitive circumstances, it is possible to catch up with South Korea in a few years. In order to accelerate the growth rate, the government should play a more active role in promoting broadband adoption and development. In addition, the threshold of high internet diffusion rate depends on the personal computers possession rate, which is over 78% in South Korea. For either

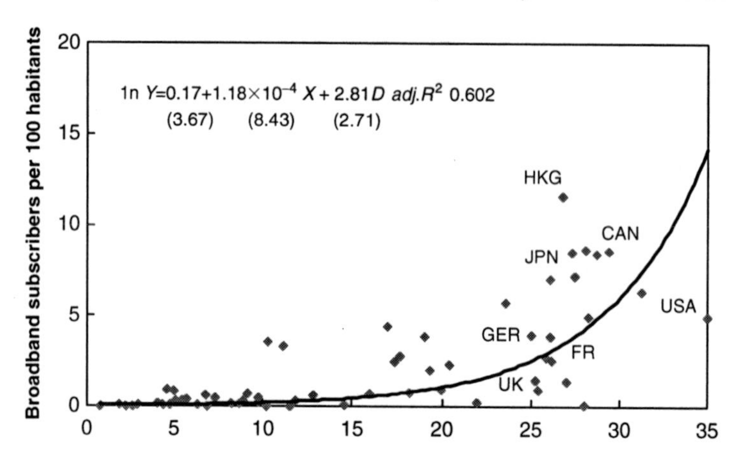

Fig. 6.18 Correlation between Broadband Subscribers per 100 Inhabitants and GNI per Capita Worldwide (2002)

[a] GNI, Gross National Income using PPP

[b] Broadband subscribers per 100 inhabitants in leading countries in 2002 are: Korea (21.3), Hongkong (14.9), Canada (11.2), Taiwan (9.4), Sweden (7.8), Netherlands (7.2), Japan (7.1), USA (6.9), Finland (5.3), Germany (3.9), France (2.4), UK (2.3), and Italy (1.5)

[c] D, Dummy variables (Korea = 1, Other countries = 0)

[d] Korea, Luxemburg and Norway are omitted from the graph but included in the trend line computation

[e] Original statistics are based on ITU [34]

dial-up, fixed broadband or wireless internet access subscribers, at least one desktop/laptop computer is necessary. Thus, in order to develop the qualified internet infrastructure further, constructing an inducing dynamism in leveraging both high rates of personal computers possession and internet access subscription should be important strategies for the government to endeavor [35].

6.4 Conclusion

6.4.1 General Summary

In light of the significance of the reconstruction of Japan's vicious cycle between non-elastic institutions and insufficient utilization of the potential benefits of ICT, this chapter analyzed diffusion, substitution and competition dynamism inside Japan's ICT market.

Prompted by the hypothetical view that recent advances in its IP mobile service deployment such as NTT DoCoMo's i-mode service can be attributed to a co-evolutionary dynamism between diffusion, substitution and competition inside its ICT market, an empirical analysis of the mechanism co-evolving this dynamism in telephony, mobile telephony and internet access markets was attempted by utilizing four types of diffusion models identical to respective diffusion, substitution and competition dynamics.

6.4.2 New Findings

6.4.2.1 Telephony Market

(a) The simple logistic model demonstrate that the potential of fixed line phone market in Japan is about 62 million, lower than that of the mobile phone since fixed line phone is usually shared by people who live together such as family members while mobile phone is always possessed by individual.
(b) However, the subscription of fixed line phone has started to decline after reaching the peak in 1996. It is considered that the complementary feature between mobile and fixed line telephone is the reason for such decrease in a matured fixed phone line market.
(c) By taking mobile and fixed line telephony as two generations of telephony with substitutability and analyzing with the choice-based substitution diffusion model, it is demonstrated that the potential of the overall telephony can be about 100 million, consisting of double-subscribers who sign up to both mobile phone and fixed line phone.
(d) Consumers' utility to adopt fixed line phone increase as time goes by before mobile phone is introduced but deceases with time after mobile phone appeared in the market. The price of monthly subscription fee of mobile phone is a negative factor for users to adopt or switch to mobile phone, moreover, the double-subscribers reveal the highest price elasticity and non subscribers seem to consider price less than those who have already subscribed to either one service. The result can be attributed to the fact that for people who have not yet adopted any telephony service, they have higher need to subscribe than those who just consider to switch.
(e) Moreover, even after mobile phone is introduced into the market, existing subscribers of fixed line phone still tend to continue such subscription as long as the price of mobile phone does not drop too much.
(f) In both mobile and fixed line telephony cases, it is observed that voice-only service is occupying less and less share of communication market while data-and-voice service is still growing. With only mobile and fixed line telephony taken into consideration, the substitution effect lead the subscription of fixed line phone to decrease, however, including the impact by the internet access demand, the fixed line phone demand will decrease less than analyzed with this model.

6.4.2.2 Mobile Telephony Market

(a) Both simple logistic model and bi-logistic model demonstrate that the potential of mobile phone market in Japan is between 83 million and 85 million, with the speed that half of the potential can have been reached within about 13 years.

(b) While the diffusion process is divided into two impulses, the bi-logistic model demonstrates that the potential of market increased from 50 million to more than 83 million after the IP mobile phone was introduced into the market.
(c) However, with the choice-based substitution diffusion model that takes the substitution of NonIP by IP mobile phone into consideration, the potential of NonIP mobile phone was originally about 56 million, but the penetration rate slowed down before reaching this level and users switching from NonIP to IP mobile phone service make the diffusion of IP mobile phone move faster than expected. The overall potential of mobile phone is still estimated as about 83 million with this model.
(d) The analysis with the choice-based substitution diffusion model demonstrates the factors that affect consumers' adoption intension. Consumers' utility of adopting NonIP mobile phone increases as time goes by before IP mobile phone is introduced. However, after IP mobile phone appeared in the market, consumers' utility of adopting NonIP decreases with time due to obsolescence. Users' utility of keeping using NonIP mobile phone increases with time but more slowly than that of adopting or switching to IP mobile. As expected, consumers are more likely to adopt IP mobile phone with reduced price and enriched functions.
(e) Finally, based on the previous analysis, analysis with logistic growth within a dynamic carrying capacity demonstrates that, with the consideration about positive impact of enriched functions on ICT development, the potential of mobile phone can actually reach to about 94 million, which is higher than the estimates with previously implemented models.
(f) Observations of the newly gained market share shows that the relatively stable market might be changed in the near future. This is an evidence of more and more competitive environment that is a positive driving force of mobile phone diffusion.

6.4.2.3 Internet Access Market

(a) The simple logistic model demonstrates that the potential of the internet access market in Japan is about 37 million, even lower than that of the fixed line phone. The reason is considered to be that internet access diffusion is limited by personal computer possession rate. Moreover, fixed line internet access can also be shared by people who live together such as family members just as in the case of fixed line phone.
(b) While the diffusion process is divided into two impulses, dial-up access and broadband access, the bi-logistic model demonstrates that the potential of mobile phone market in Japan increased by about 12 million contributed by broadband internet access service. The impulse cause by broadband internet access grows at a speed faster than the first impulse of dial-up access. Such a difference in diffusion speed is attributed to the always-connected feature and the high access speed of broadband that is more appealing to the internet users.

(c) Moreover, with the choice-based substitution diffusion model that takes the substitution of dial-up access by broadband internet access into consideration, the potential of internet access was originally only about 8 million, but after broadband access is introduced, both subscription of dial-up access and broadband internet access increase dramatically.
(d) The analysis with the choice-based substitution diffusion model demonstrates that both time and price affect consumers' adoption intension. Time also represents the degree of maturity of internet technology as well as the bandwidth, thus it is rational so have the result that consumers' utility increases with time. Consumers' utility of using broadband access increases faster with time than that of using the dial-up service.
(e) However, the case of internet is different from that of the telephony because users' utility of adopting dial-up access increases with time no matter before or after broadband access is introduced. This reason is considered to be that people still sign up to dial-up service while broadband has already been introduced into the market because they consider themselves as non-frequent-users of the internet. Therefore, they can still be satisfied with the limited speed offered by dial-up access.
(f) The penetration rate of dial-up access that has started to slow down since 2002 is expected to decline rapidly during the coming years. Users switching from dial-up access to broadband access make the diffusion of broadband access move faster than expected. The overall potential of the internet access is estimated as about 40 million with the choice-based substitution diffusion model.
(g) Competition is expected to be the most important driving force to fill up the gap between broadband market development in Japan and Korea.

6.4.2.4 Diffusion, Substitution and Competition Dynamism Inside the Transitional Market in Telephony, Mobile and Internet

All these findings obtained in the substitution dynamism in the markets of telephony (fixed line to mobile), mobile telephony (NonIP mobile to IP mobile) and internet access (dial-up to broadband) demonstrate a noting co-evolutionary dynamism between diffusion, substitution and competition emerging inside the Japanese ICT market in transition. Key factors governing this dynamism are identified as ICT innovation, enriched functions, reduced price and competitive environment.

6.4.3 Policy Implications

Given the foregoing dynamism can be the south of Japan's noting advances in IP mobile servile deployment, despite a lack of institutional elasticity, systems approach in stimulating a co-evolution between ICT innovation, diffusion, substitution and competition with the special attention to the following policies would be essential:

(a) With IP mobile phone as the mainstream of telephony market, it is expected that everyone can have a ubiquitous information receiver. Being able to communicate with others via not only voice but also data will definitely increase the mobility and hence improve the efficiency of necessary communication. Deregulations to promote competition are furthering smoothly in most advanced countries, and the government should continue to make effort to keep such competitive environment rolling. Moreover, the possibility of mobile phone may be extended from only e-mail receiver to more practical internet terminal. Making mobile internet a required function for convenient life will increase its penetration rate more efficiently.
(b) With the substitution by mobile phone, fixed line phone seems to be considered as a sunset industry. However, judging from the increasing demand for data-and-voice service, it is possible that the demand for fixed line phone will increase again if the demand for the internet is sufficient.
(c) Currently the broadband access still depends on the fixed line phone infrastructure, so the government should take such impact into consideration when deciding to continue expanding fixed line phone infrastructure development or not.

6.4.4 Future Works

(a) In this research, the empirical diffusion and substitution analysis of ICT market is focused on Japan's market. However, given the significance of such analysis done with data of more countries, similar analysis should be conducted for other countries.
(b) In the part of telephony market analysis, it s noted that the ratio of double-subscribers are so difficult to capture that we can only facilitate the result of annual investigation. However, with the subscription rate reported monthly, double-subscriber ratio should also be collected on a monthly base if possible.
(c) The personal computer market is expected to reveal similar characteristics of diffusion and substitution. However, unlike services that users have to subscribe to, personal computers are hardware that consumers buy and go. It is more difficult to keep track in the real possession rate of personal computer rate. Similar analysis should be conducted with sufficient data of personal computers if possible.
(d) Based on the result of personal computers diffusion analysis, the correlation between personal computers penetration rate and the internet access penetration rate can be analyzed to decompose their interaction, including the restriction that personal computers possession rate might have on the internet access rate and the contribution that the internet access might have on personal computers purchasing.

(e) Finally, the whole ICT market including the service section and the hardware section can be analyzed from a comprehensive level. It is expected that the ICT policies can be made from a more comprehensive view with such analysis conducted well.

References

1. R. Kondo and C. Watanabe, The virtuous cycle between institutional elasticity, IT advancement and sustainable growth: can Japan survive in a information society? Technology in Society 25, No. 3 (2003) 319–335
2. P. Schreyer, The impact of information and communication technology on output growth, OECD STI Working Paper, Organization for Economic Corporation and Development, Paris, 2000
3. OECD, OECD Growth Project Report, OECD, Paris, 2001
4. International Telecommunication Union (ITU), ITU Telecommunication Indicators Database, ITU, Geneva, 2003
5. World Bank, World Development Indicator, World Bank, Washington, DC, 2003
6. InfoCom Research, Inc., Mobile Comminication Services, Information and Communications in Japan, Tokyo, 2002
7. NTT Mobile Communications Network, Inc., NTT DoCoMo's Vision Towards 2010, NTT Mobile Communications Network, Inc., Tokyo, 1999
8. J.W. Woodlock and L.A. Fourt, Early prediction of market success for grocery products, Journal of Marketing 25 (1960) 31–38
9. E. Mansfield, Technical change and the rate of imitation. Econometrica 29 (1961) 741–766
10. F.M. Bass, A new product growth model for consumer durables, Management Science 15, No. 5 (1969) 215–227
11. C. Easingwood, V. Mahajan and E. Muller, A nonuniform influence innovation diffuison model of new product acceptance, Marketing Science 2, No. 3 (1983) 273–295
12. J.A. Norton and F.M. Bass, A diffusion theory model of adoption and substitution for successive generations of high-technology products, Management Science 33, No. 9 (1987) 1069–1086
13. V. Mahajan and E. Muller, Timing, diffusion and substitution of successive generations of technological innovations: The IBM Mainframe Case, Technological Forecasting and Social Change 51 (1996) 109–132
14. D.B. Jun and Y.S. Park, A choice-based diffusion model for multiple generations of products, Technological Forecasting and Social Change 61 (1999) 45–58
15. R.A. Peterson and V. Mahajan, Multi-product growth models, Research in Market Forecasting (JAI Press, Greenwich, 1978)
16. P. Parker and H. Gatignon, Specifying competitive effects in diffusion models: an empirical analysis, International Journal of Research in Marketing 11 (1994) 17–39
17. R.C. Rao and F.M. Bass, Competition, strategy and price dynamics: a theoretical and empirical investigation, Journal of Marketing Research 22 (1995) 283–296
18. E. Dockner and S. Jorgensen, Optimal pricing strategies for new products in dynamic oligopolies, Marketing Science 7 (1998) 315–334
19. J.T. Teng and G.L. Thompson, Oligopoly models for optimal advertising when production costs obey a learning curve, Management Science 29, No. 9 (1983) 1067–1101
20. D. Horsky and K. Mate, Dynamic advertising strategies of competing durable good producers, Marketing Science 7, No. 4 (1988) 356–367
21. G.L. Thompson and J.T. Teng, Optimal pricing and advertising policies for new product oligopoly models, Marketing Science 3, No. 2 (1984) 148–168
22. V. Mahajan, S. Sharma and R.D. Buzzell, Assessing the impact of competitive entry on market expansion and incumbent sales, Journal of Marketing 57 (1993) 39–52

23. P.D. Allison, Discrete-time methods for the analysis of event histories, Sociological Methodology Jossey-Bass San Francisco, 1982
24. S.S. Oren and M.H. Rothkopf, A market dynamics model for new industrial products and its application, Marketing Science 3 (3) (1984) 247–265
25. P. Hedstrom, Contagious collectivities: on the spatial diffusion of Swedish Trade Unions, 1890–1940, American Journal of Sociology 99, No. 5 (1994) 1157–1179
26. N. Cadima and P.P. Barros, The impact of mobile phone diffusion on the fixed-link network, Discussion Paper DP2598, Centre for Economic Policy Research, London, 2000
27. H. Gruber and F. Verboven, The evolution of markets under entry and standards regulation – the case of global mobile telecommunications, International Journal of Industrial Organization 19, No. 7 (2001) 1189–1212
28. M. Rodini, M.R. Ward and G.A. Woroch, Going mobile: substituability between fixed and mobile access, Telecommunications Policy 27 (2003) 457–476
29. D.B. Jun, S.K. Kim, Y.S. Park, M.H. Park and A.R. Wilson, Forecasting telecommunication service subscribers in substitutive and competitive environments, International Journal of Forecasting 18 (2002) 561–581
30. M.N. Sharif and K. Ramanathan, Binominal innovation diffusion models with dynamic potential adopter population, Technological Forecasting and Social Change 20 (1981) 63–87
31. Telecommunications Carriers Association (TCA) Japan, http://www.tca.org.jp/ (TCA, Tokyo, 2004)
32. Ministry of Public Management, Home Affairs, Posts and Telecommunications (MPHPT), Fixed Line Subscription Report, Information and Telecommunicatons Statistics Database, http://www.johotsusintokei.soumu.go.jp/index.html (MPHPT, Tokyo, 2003)
33. Japan Information Processing Development Corporation (JIPDC), IT White Book (JIPDC, Tokyo, 2003)
34. International Telecommunication Union (ITU), Birth of Broadband, ITU Internet Report (ITU, Geneva, 2003)
35. J. Hamanaka and C. Watanabe, Comparative analysis of institutional elasticity for maximizing the effect of industrial technology policy – a cross-country comparison of the diffusion and adaptation process of IT, Report of NEDO Research Project, New Energy and Industrial Technology Development Organization, Tokyo, 2003

Chapter 7
The Co-Evolution Process of Technological Innovation: *An Empirical Study of Mobile Phone Vendors and Telecommunication Service Operators in Japan*

Abstract While the development of information and communication technology (ICT) is usually measured by quantitative indices such as penetration rate, the qualities and characteristics of all countries can be different even if their achievements seem to be similar judging from quantitative standards. The mobile phone market represents the kind of market that should be analyzed from both quantitative and qualitative perspectives.

In this chapter, the empirical analysis focuses on the mobile phone market in Japan, which has achieved a high internet access rate through mobile phone as well as a significant array of applications. The first analysis, by classifying the existing handset models, shows that high-end handsets occupy the largest share in Japan's market, and the ratio is much higher than the worldwide average level. Further analysis reveals that most domestic handset vendors offer order-made models to maintain high quality and to meet the specific demands of each telecommunication service operators. As a result, global handset vendors find it difficult to enter the co-evolution cycle by simply offering global models to the service operators. Moreover, although the tight tie-up between service operators and handset vendors may be criticized as a conservative or closed relationship, consumers' strong consciousness toward high quality and innovative functions creates a complex demand driven marketplace. In this environment, service operators and handset vendors' commitment to quality enables the virtuous cycle of technological innovation to "roll" along smoothly.

Reprinted from *Technology in Society 29*, No. 1, C. Chen, C. Watanabe and C. Griffy-Brown, The Co-evolution Process of Technological Innovation: An Empirical Study of Mobile Phone Vendors and Telecommunication Service Operators in Japan, pages: 1–22, copyright (2007), with permission from Elsevier.

C. Watanabe, *Managing Innovation in Japan: The Role Institutions Play in Helping or Hindering how Companies Develop Technology*, DOI: 10.1007/978-3-540-89272-4_7,

7.1 Introduction

7.1.1 Background

Rapid growth of information and communication technology (ICT) has attracted much attention in both the business and academic fields. Among all aspects of ICT, mobile phone and the internet are two sectors with a high growth rate. As the International Telecommunication Union (ITU) points out, "*virtually all of the growth in the global telecoms sector over the past decade has come from mobile communications and the internet*" [1]. The total number of mobile subscribers all over the world has risen over 1,300 million, with the penetration rate up to 24 users per 100 nations as demonstrated in Fig. 7.1.

However, after these technological innovations spread throughout the world, the maturity of a market can no longer be judged only by the penetration rate. The degree of market development should be evaluated not only by the quantity of use but also by the quality of use. Taking the overlapping technology of the mobile phone and the internet as an example, mobile internet broadens the possibility of mobile phone functions and allows more advanced applications. Among all the countries with a high-level penetration rate of mobile phone, Japan and the Republic of Korea were the first two nations to launch third generation mobile networks commercially. According to 2005 White Paper of Information and Communications in Japan [2], more than 90% of mobile phone service subscribers in Japan can access to the internet through the handsets that is the world's highest rate, much higher than the 33.5% level in the US as illustrated in Fig. 7.2. This can be partially attributed to the reality that subscribers enjoy much more functions through mobile internet access [2]. Furthermore, since ITU also predicted in its Asia-Pacific Telecommunication Indicators

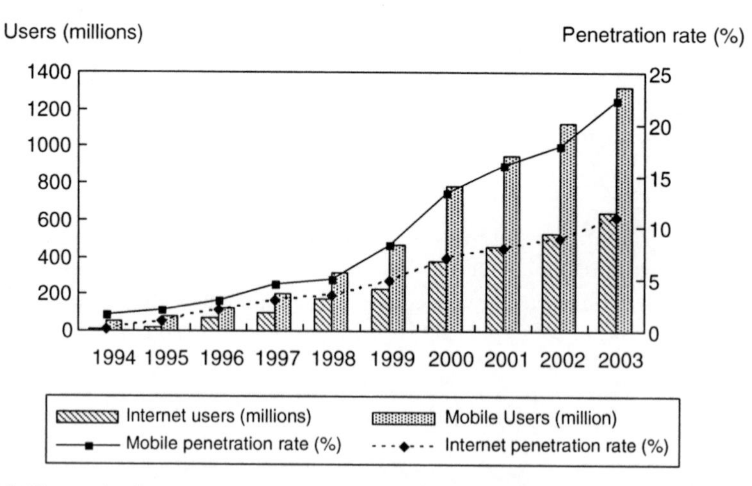

Fig. 7.1 The number/penetration rate of mobile subscribers and internet users in the world (1994–2003)
Source: Internet Reports 2004: The Portable Internet [1]

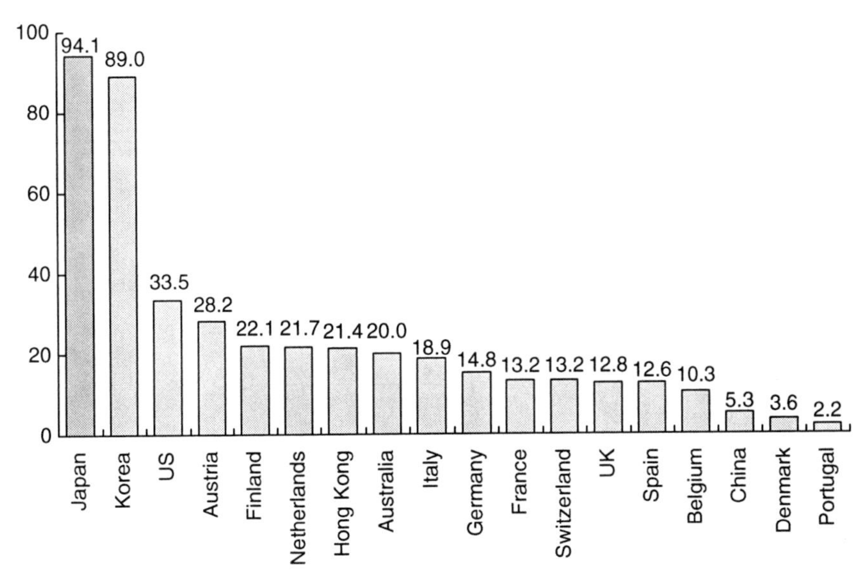

Fig. 7.2 Ratio of mobile internet subscribers among mobile phone subscribers in leading countries (Sep. 2004)
Source: 2005 White Paper of Information and Communications in Japan [2]

2002 that the global telecommunications epicenter is shifting from North America and Western Europe to Asia-Pacific [3], Japan, a developed market of mobile phone in Asia, can be considered as an important model suggesting how other Asian markets might evolve. In 2000, American journalists also suggested in Business Week that the wireless internet service that has become so popular among the Japanese could certainly spread around the world [4].

The handsets in Japan are equipped with various functions such as an internet browser, music player, mobile phone camera, game applications and so on. All the functions require the contribution of both service providers and handset vendors. Consequently, the distinctive features of the Japanese market are that the service operators and all handset vendors work closely together. One good example is the mobile internet functionality. In addition to the advanced service offered by the service operators that utilizes mobile internet, handset vendors also provide the subscribers in Japan with well-developed handsets equipped with an internet browsing application. Taking the digital camera on the handset as another example, over three fourths of mobile subscribers in Japan can take pictures with the mobile handsets instead of using the handsets only for talking in 2005 [2]. Figure 7.3 shows the number of mobile subscribers and the ratio of handsets with camera features over the period 2001–2005 which demonstrate a conspicuous increase in the ratio particularly from 2003.

Such a rapid diffusion of camera-attached mobile handsets demonstrates the rapid growth of mobile phone driven innovation. Japan's demanding consumers induce new functionality and they can also learn promptly to use new features leading

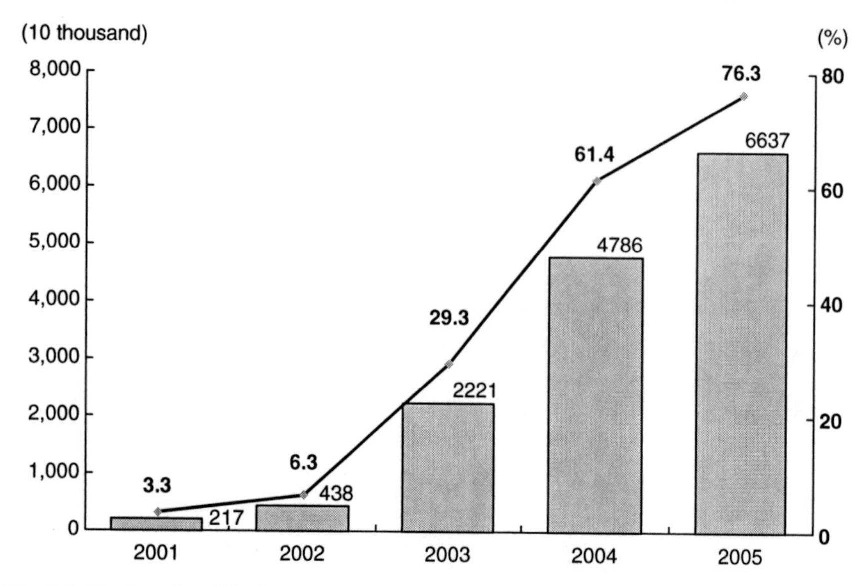

Fig. 7.3 Number of mobile phone subscribers and the ratio of mobile handset with camera feature in Japan (2001–2005)
Source: 2005 White Paper of Information and Communications in Japan [2]

to the high-end handset dominated market. Furthermore, new functionality also promotes the learning effect of the market in Japan in a long-term perspective. Consequently, the close relationship between service operators and handset vendors enables a continuous supply of mobile phones with high-quality/new functionality to satisfy consumers' demand. At the same time, consumers also learn more promptly to adapt themselves to the new functionality and then become more demanding. Such a virtuous cycle has been involving the close cooperation of consumers, handset vendors and service operators drives the co-evolution of demand side and supply side in Japan.

7.1.2 Hypotheses

The foregoing observations prompt us to consider the following hypothetical views with respect to the systems dimension of the sources that enabled Japan's conspicuous advancement of mobile phone services:

(a) Customers' demand for new functionality and high learning capability to use new features construct a market dominated by high-end handsets.
(b) In order to provide the market with high-quality handsets with new functionality all the time, service operators and handset vendors need to construct a closer relationship which might consequently mold Japan on a closed but independent market from others.

(c) The customers' high demand for mobile phones with high-quality/new functionality and close relationship between service operators and handset vendors construct a virtuous cycle that drives the co-evolution between demand side and supply side in Japan.

7.1.3 Existing Works

Many studies have considered the diffusion of technological innovation from the quantitative perspective. Woodlock and Fourt [5], Mansfield [6], Bass [7], Easingwood, Mahajan and Muller [8], etc. analyzed the well-known first-purchase diffusion of new products in marketing by emphasizing the consumers' imitation and adoption behavior of product and constructed models dependent on the variable "time." Jun and Park [9] and Jun et al. [10] proposed a diffusion model for multiple generations and included other factors such as the price of products which reveals how price may affect the diffusion curve by affecting customers' choice. Although Yamada and Furukawa [11] demonstrated that almost all Japan's home electronic appliances were diffused at a slower rate than other markets, the advancement of mobile phone market seems to be an exception. Iimi [12] suggested that the mobile phone market in Japan is highly product-differentiated as well as service-differentiated so that conventional concerns such as the size of network are no longer decisive factors in choosing a mobile phone carrier. Iimi's empirical analysis focuses on the billing plan instead of the technical specification of the handsets. Based on Jun and Park's empirical analysis [9] on Korean telecommunications market, Chen and Watanabe [13] further developed a model to take "function" as a variable that affects the diffusion curve by focusing on Japan's mobile phone market. In this empirical analysis by Chen and Watanabe, mobile phones without internet service as the first generation and those without internet service as the second. However, due to the limitation of the numerical model, even if it is identified that "price" and "new features" of mobile phone are important factors that pushed the rapid growth of more advanced mobile phone penetration rate in Japan, it is not specified which functions determined the advancement of mobile phone market. In addition, since almost all mobile phones are equipped with mobile internet services, a further classification approach based on a precise definition of technical specifications (functions) is required to understand the segmentation of this market.

Regarding the supply side factors that affect the progress of mobile phone market, Kondo and Watanabe [14] suggest that the non-elastic institutions existing in Japan might cause Japan to lose its international competitiveness but the case of mobile phone market is an exception. Nagamachi [15] points out that Japan's manufacturers (including mobile handset vendors Sharp, Sanyo, and Matsushita) keep making efforts to design products base on Kansei engineering which emphasizes consumer-oriented technology and take consumers' satisfaction as the most important goal. This unique insistence of Japanese manufacturers reveals their high concerns about consumers' satisfaction and the demand driven nature of this sector. Kodama [16] analyzed the community-based firm structure of NTT DoCoMo and explained the way NTT DoCoMo opted to alter employee consciousness, vitalize organizational morale, entrench the new NTT "Phoenix" brand (video conferencing system) in the Japanese market, and create an emergent new video conferencing market. Funk [17] also emphasizes the importance of firms that provide services, content and technologies in the mobile phone market. Jonsson and Miyazaki [18] compared the 3G mobile phone strategies of top two service operators NTT DoCoMo and au KDDI and mentioned NTT DoCoMo's high-cost and high-quality strategies earns support in Japan but lead to failure in the European market. Iimi [19] identifies that almost no effect of price competition across regions and the incumbent carriers keeps essentially related technology monopolized even after privatization and liberalization. All the studies above suggest there are unique institutions in Japan, but most of them focus on the analysis of strategies of a singular service operator or manufacturer. No attention is paid to the relationship between the service operators and handset vendors, i.e. the characteristics of the supply side in the market and its intimate relationship with the demand side.

7.1.4 Prime Objectives of the Investigation

While Japan's rapid advancement of mobile phone service is expected to be an important model for similar advancement in other countries, applying solely the experience of Japan to other markets does not promise the same level of success due to various distinct market attributes. First, even if it is noticed that Japan has achieved a relatively high performance in the diffusion and innovation of mobile phone, the attributes of the market that made this possible are unclear. How to identify the quality and level of this market requires careful examination. Next, what are the factors that determine such an advanced market? Both the consumer side and supply side should be taken into consideration. Is R&D of the service operators and handset vendors or the consumers' demanding attitude toward new features fuelling the high growth rate of new mobile services, such as mobile internet, and innovative handset features, such as the mobile camera? Last, the mechanism that continues to drive the success of the mobile phone market in Japan should be the key factor for elucidating similar mobile phone markets. Demonstration of the three hypothetical views mentioned above would provide significant insight into these questions.

Therefore, this research attempts to demonstrate these hypotheses in the following ways. First, this study uses exact indicators to define "advanced" mobile features and show how advanced the mobile phone market in Japan is compared with the average level around the world. Further analysis by utilizing a learning curve model is conducted to analyze the trend in the learning curve coefficient in order to identify how the progress of new functions affected the mobile phone market in Japan. Second, in addition to identifying the uniqueness and quality of Japan's mobile phone market, the extremely close and mutually dependent relationship between the service operators and handset vendors in Japan is identified. Based on these analyses, the reason why the worldwide handset vendors are unable to perform in Japan as they do in other countries is explained. Finally, a co-evolution mechanism between the consumers and the supply side, including handset vendors and service operators, is elucidated in order to point out the barriers for worldwide handset vendors in penetrating such a well-defined market. This research also aims at explaining how the Japanese "inflexible" institutional system can drive the growth of penetration rate and enable the high-quality/advanced-features of the mobile phone market while impeding the Japanese service operators and handset vendors from succeeding in foreign markets.

7.1.5 Structure of the Chapter

Section 7.2 demonstrates how the marketplace is dominated by high-end handsets and how it affects the learning curve of the market. Section 7.3 demonstrates how the service operators and handset vendors construct a close relationship to satisfy the consumers' demand and how this relationship molds this market. Section 7.4 demonstrates a virtuous cycle that drives a co-evolution between demand side and supply side in Japan due to its unique institutional system. Section 7.5 briefly summarizes new findings, policy implications and points toward future work in this area.

7.2 A Market Dominated by High-End Handsets Due to High Learning Effect

7.2.1 Methodology

7.2.1.1 Analysis of the Degree of High-End Handsets

In order to demonstrate that the mobile phone market in Japan is much more advanced than the average level in other countries, it is necessary to quantify the features of handsets by classifying the handsets currently available in Japan's mobile phone market into three segments: high-end, midrange and entry-level. Since the handset vendors and operators keep releasing new models each season, the first step of this analysis is to clarify the range of handset models to be analyzed. The target handset models are those listed in the catalogs and websites of the operators in September 2005, and the specification information is collected from the catalogs and websites [19–21].

Next, the criteria to classify the models into three segments are fixed. In order to keep the consistency of classification standard with other existing analysis, the classification methodology is based on the aspects suggested by Slawsby and Leibovitch [22]. However, since "form factors," such as small monoblock (not-foldable style)

Table 7.1 Criteria to classify high-end, midrange and entry-level handset

Feature	High-end	Midrange	Entry-level
Application processor	Integrated	Discrete	None
Memory	<32 MB	<16 MB	<4 MB
Display	Color display, 240×320 pixels or less	Color display, 100×100 pixels or less	Monochrome display
Wireless generation	2.5G or 3G+	Mostly 2.5G	Mostly2G
Expansion	Yes	None	None
Camera	Standard, 1.5 MP or more	Standard, <1.5 MP	None
Applications	Full multimedia stream and content	Simple web-browsing and downloading games	Phonebook and text-messages

Source: Slawsby and Leibovitch, worldwide mobile phone 2005–2008 forecast by feature tier: a feature-rich future [22]

and clamshell (the foldable style), depends on the preference of usage and does not necessarily mean which one is more advanced than another, it is not included in the indicators used in this research. Since one "operating system" does not show clear superiority than another, this item is also neglected. All features taken into account are listed in Table 7.1. For each feature classified as high-end, the handset is credited with three points, two points for each feature classified as midrange, and one point for each feature classified as entry-level. Wireless generation is the exception since 2.5G handsets may be high-end or midrange. However, stricter criteria are preferred for high-end handset, so only 3G handsets obtain three points in "wireless generation" feature in this research. After crediting the points for the seven features of all handset models, the average point of each model is computed. One handset model is classified as high-end if the average point is 2.5 or more, midrange if 1.5 or more, and entry-level if less than 1.5.

Based on the classification analysis results, the total number of high-end, midrange and entry-level models of each handset vendor and each operator is computed. By comparing the total score of the credits, it can be observed which segment each handset vendor and operator is focusing on.

7.2.1.2 Analysis of the Learning Effects in the Mobile Phone Market in Japan

In order to assess not only the innovation mechanism of the supply side of mobile phone industry in Japan but also the learning capability of customers toward new functions and technology, the learning effect is analyzed based on a learning curve where *the average price index of mobile phone (P)* is influenced by *the number of mobile phone subscribers (N)* with a dynamic elasticity (learning curve coefficient) as a function of the determinants *time (t)*. The number of mobile phone users is based on the monthly reports released by the Telecommunication Carriers Association in 2005 [23], and the price index of mobile phone is collected from the monthly reports of price index published by the Bank of Japan (1996–2004) [24]. The determinant *time (t)* represents the maturity of technological innovation of the supply side while *the number of mobile phone subscribers (N)* corresponds to the population involved in the learning process.

7.2.2 Results and Discussion

7.2.2.1 High-End Oriented Vendor Structure in Japan's Mobile Phone Market

By listing all target models and completing the classification, the results are shown in Tables 7.2 and 7.3 by handset vendor and service operator, respectively.[1]

The result of the classification analysis demonstrates that all Japan's handset models are classified in either high-end or midrange, and high-end handset models share an extremely high ratio. According to the news release of IDC Japan in 2005,

Table 7.2 Classification of handset models by handset vendor

Number of models	Segment		
Handset vendor	High-end	Midrange	Total
Casio		1	1
Fujitsu	5	4	9
Kyocera		1	1
Mitsubishi	2	4	6
NEC	6	3	9
Nokia		1	1
Panasonic	4	5	9
Sanyo		4	4
Sharp	5	8	13
SonyEricsson	8	6	14
Toshiba		5	5
Total	30	42	72

Table 7.3 Classification of handset models by service operator

Number of models	Segment		
Service operator	High-end	Midrange	Total
NTT DoCoMo total	19	19	38
Au Kddi total	8	12	20
Vodafone total	3	11	14
Total	30	42	72

[1] Since Tuka and NTT DoCoMo (PHS) as a whole occupy less than 5% of the market, they are neglected in the table of handset vendors.

the top four vendors in 2004 are NEC, Sharp, Panasonic and SonyEricsson [25]. As listed in Table 7.2, these top four handset vendors provide the service operators with more high-end handset models. On the other hand, judging from the segment analysis of each service operator, it is shown that the top service operator NTT DoCoMo, which shares about 55% of the market, provides consumers much more high-end models than other service operators. Half of NTT DoCoMo's models belong to high-end while high-end models share only 20% of models provided by Vodafone, whose market share is the lowest among the three major service operators. These analyses demonstrate that, compared with the ratio of the worldwide market, Japan's mobile handset market is obviously dominated by the high-end handsets. As compared in Fig. 7.4, while worldwide high-end ratio remained 7.8% in 2004, Japan demonstrates conspicuously high-level as 41.7% in 2005.

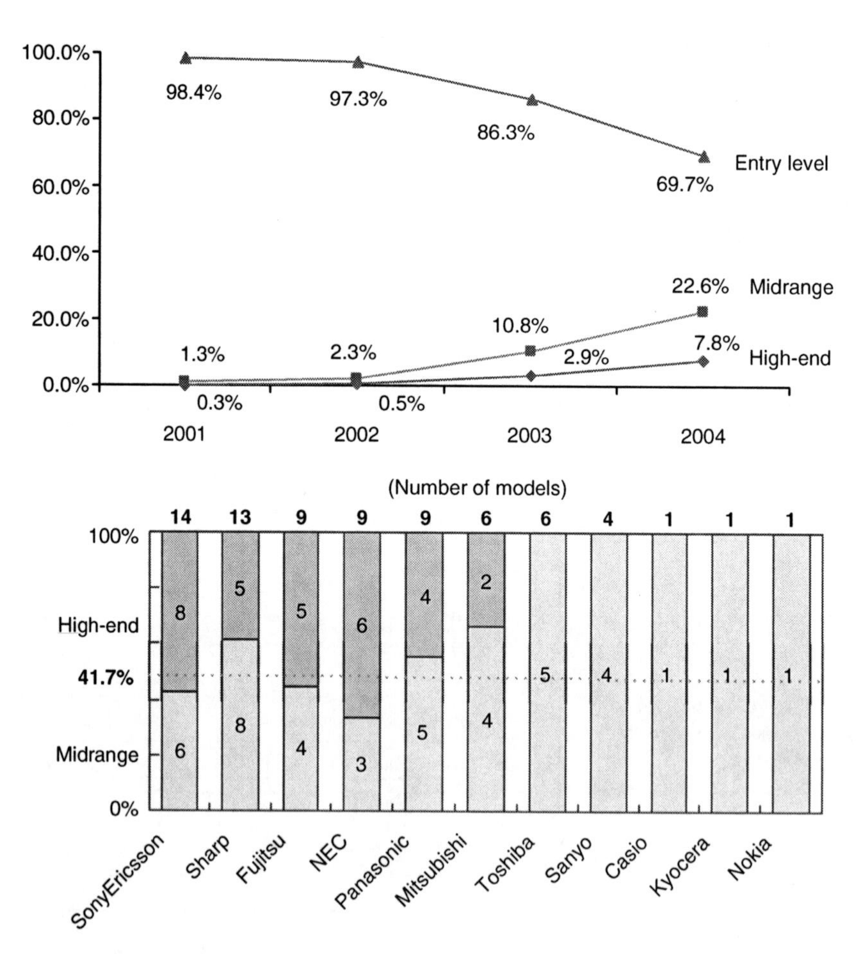

Fig. 7.4 Comparison of the ratio of high-end, midrange and entry-level handset ratios in the worldwide market (2001–2004) and in Japan (Sep. 2005)
Source: IDC [25] and empirical analysis of this research

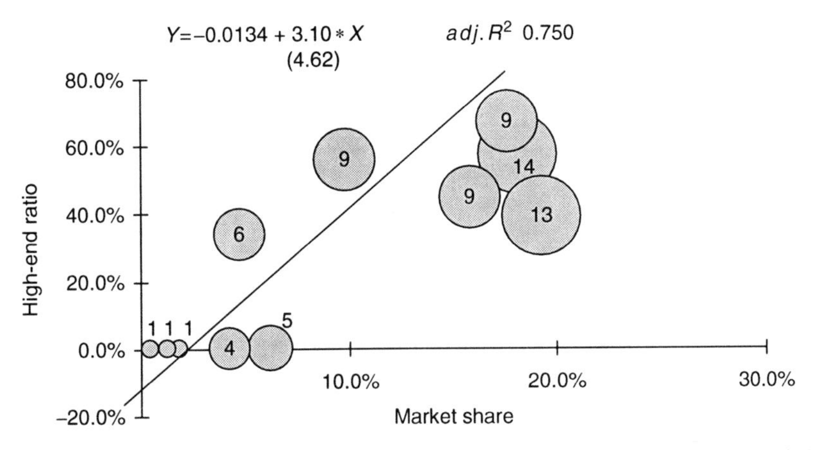

Fig. 7.5 Correlation between market share and high-end ration in Japan's handset vendors (2004)

Figure 7.5 demonstrates a positive correlation between market share and high-end ratio in Japan's mobile handset vendors. All demonstrates Japan's high-end oriented vendor structure.

7.2.2.2 Analysis of Learning Effect of Mobile Phone Market in Japan

According to the regression analysis, at the early stage of the mobile phone market, the mobile phone market was still monopolized and the price did not actually reveal the interaction of the supply side and the consumer side. Therefore, in order to maintain the statistical significance, the time-series data before 1997 is omitted from this analysis. The parameters of the learning effect model are estimated by applying the empirical data from Jan. 1997 to Feb. 2002; with the estimated model, the learning coefficient from Jan. 1997 to Jun. 2006 is computed.

The result of the empirical analysis of the learning effect model with dynamic learning curve coefficient identifies with statistical significance the correlation between the cumulated number of mobile phone subscribers and the price of mobile phones in Japan (Jan. 1997–Feb. 2002) as follows:

$$\ln P = \underset{(42.66)}{10.03} - (\underset{(24.74)}{5.76 \times 10^{-1}} - \underset{(7.97),}{1.78 \times 10^{-5} t^2} + \underset{(6.07)}{1.80 \times 10^{-7} t^3}) \ln N \quad adj.R^2 = 0.986$$

where P, the price index of handsets, t, time and N, the cumulative number of mobile subscribers.

This result suggests that the learning curve coefficient of Japan's mobile phone market λ can be depicted as follows as a function of time t:

$$\lambda = 5.76 \times 10^{-1} - 1.78 \times 10^{-5} t^2 + 1.80 \times 10^{-7} t^3.$$

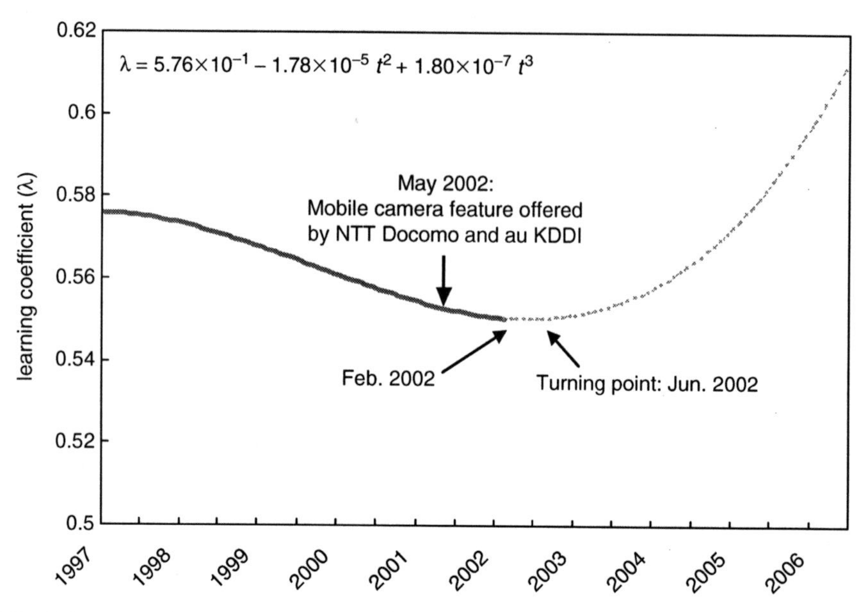

Fig. 7.6 Estimates[a] and extended estimates[b] of the learning curve coefficient in Japan's mobile phone market (Jan. 1997–Jun. 2006)

[a] Jan. 1997–Feb. 2002

[b] Mar. 2002–Jun. 2006

[c] Original statistics are based on monthly reports by Telecommunication Carriers Association (TCA), Japan (2005) and price index reports of Japan Bank (2004)

As illustrated in the learning coefficient curve (Jan. 1997–Jun. 2006) in Fig. 7.6, it is observed that after the mobile phone appeared in the market, the learning coefficient kept decreasing slowly as has been generally observed in the initial stage of the diffusion of new innovations. Existing empirical analysis also suggests that the learning coefficient of the market tends to drop due to technology obsolescence. However, the inflection point of the learning curve coefficient in the mobile phone market in Japan lies on the point representing Jun. 2002. Judging from the list of service/functions in Table 7.4, the curve turned upward after NTT DoCoMo and au KDDI started to offer handsets with the "mobile camera" feature in May 2002. Mobile handsets with camera features pulled up the learning coefficient and other functionalities accelerated the enhancement of learning curve coefficient.

In fact, J-phone (now Vodafone) has supplied handsets with the mobile camera feature since Dec. 2000. Unfortunately, its market share was not high enough, so the effect was relatively minor. On the other hand, the effect of new features was significant when the top two service operators, NTT DoCoMo and au KDDI released handsets with the same feature. This observation demonstrates strong learning effects as a consequence of the intensive interaction between high-end oriented supply and demanding consumers leading further qualified functions in Japan's mobile phone market. These learning effects again support the perspective of the enormous power of service operators, instead of handset manufactures, in the unique mobile market of Japan.

Table 7.4 New services and functions introduced by service operators (1999–2004)

Time	New service/function	Service operator
Feb 1999	i-mode (e-mail)	NTT DoCoMo
Apr 1999	Ezweb (e-mail)	au KDDI
Dec 1999	J-sky (e-mail)	J-phone[a]
Jan 2000	DoCoNavi (navigation)	NTT DoCoMo
May 2000	J-navi (navigation)	J-phone[a]
Nov 2000	J-pic-mail (camera)	J-phone[a]
Jan 2001	i-appli (Java)	NTT DoCoMo
Jun 2001	J-Appli (Java)	J-phone[a]
Jul 2001	Ezplus (Java)	au KDDI
Nov 2001	i-motion (video clip)	NTT DoCoMo
Dec 2001	Ezmovie (video clip)	au KDDI
Dec 2001	Eznavi (navigation)	au KDDI
May 2002	i-shot (camera)	NTT DoCoMo
May 2002	au-shot (camera)	au KDDI
Dec 2002	J-movie (video clip)	J-phone[a]
Dec 2003	Movie TV	Vodafone
Dec 2004	Music downloading capable handset	NTT DoCoMo
Dec 2004	Music downloading service	au KDDI
Dec 2004	3G mobile content downloading	Vodafone

Sources: IT White Book 2003 [26] and service operators' website [19–21].
[a]J-phone was merged into Vodafone Group and renamed as Vodafone KK in Sep. 2003

7.3 A Market Dominated by Order-Made Models Due to Close Cooperation Between Vendors and Operators

7.3.1 Methodology

In order to elucidate the relationship between domestic handset vendors and service operators in Japan, it is necessary to clarify the pattern of how handset vendors provide handset models to the service operators. This analysis aims at classifying the handset models into three types: global, localized and customized model. Here "*customized model*" means that the handset vendor provides that handset model only to one service operator in the same country. "*Regional model*" means that the handset vendor provides very similar handset models to different service operators in the same country. "*Global model*" means that the handset vendor provides that handset model to different service operators in different countries.

The target handset models are those listed in the catalogs and websites of the handset vendors in October 2005; the spec information and photos of those handsets are collected from the catalogs and websites of NEC, Sharp, Panasonic, Fujitsu, Mitsubishi, Sanyo, Toshiba, SonyEricsson in September 2005 [27–34]. After the information of each model is collected, all the models are listed by handset vendor and then by service operator. By comparing the specifications and style of the same handset vendor, it can be observed whether the same or similar model is provided to different service operators. If the model is provided to only one service operator, it is categorized as "*customized*" model. If it is provided to multiple service operators in Japan by only minor changes, it is considered as "*regional*" model. If it is also provided to other service operators in other countries, then it is classified as "*global*" model.

7.3.2 Results and Discussion

7.3.2.1 Japan's "*Customized*" Model Oriented Market

The models analyzed are listed in Fig. 7.7 by handset vendor and then by service operator. Some handset vendors, such as Panasonic and Fujitsu, only provide handsets to a specific service operator, so the models are clearly classified in "*customized*" models. Mitsubishi once provided handsets to two service operators, but tend to concentrate on one service operator after 3G models become the main stream in the market. Other handset vendors, including Sanyo, Toshiba, and SonyEricsson, provide handsets to multiple service operators. However, by comparing the specification and style of the models, it is observed that these handset vendors provide different models to different service operators. Only N900iG and 802N made by NEC as well as 802SH and SH506iC made by Sharp provide similar models to different service operators and are classified in "*regional*" models. Other models made by Sharp and NEC except these two pairs of models are, similar to Sanyo, Toshiba, SonyEricsson and Mitsubishi, classified in "*customized*" models.

Based on these classifications, Fig. 7.8 listed the number of handset models as handset vendor and service operator by classifying "*customized,*" "*global*" and "*regional*" models. This demonstrates that 93% of the models in September 2005 can be classified in "*customized*" models. This analysis demonstrates the relationship between handset vendors and service operators in Japan exists in the order-made pattern of cooperation resulting in depending on "*customized*" model. It is not an open environment for a handset vendor to develop one model and simply provide it to different service operators. It is also observed that NTT DoCoMo, with its longest history of mobile phone business in Japan, demonstrates the highest ratio of customized and high-end handsets. On the other hand, Vodafone, the latest comer among the top three service operators, demonstrates the lowest ratio of customized and high-end handsets. This is considered as one of the differences of traditional Japanese service operator and global-business oriented service operator.

7.3.2.2 Institutional Sources Molding Japan's Market on "*Customized*" Model

In Japan, the mobile handsets are sold with both the name of the handset vendor and the service operator on them.[2] Therefore, from the service operators' perspective,

[2] This system is expected to changed from the autumn 2006 by the launching of "number portability service".

	NEC		SHARP		Panasonic	Fujitsu
	DoCoMo	Vodafone	DoCoMo	Vodafone	DoCoMo	DoCoMo
3G/2.5G	N901is		SH700iS	703SH	P901iS	F901iS
	N901iC		SH901iS	903SH	P700i	F700iS
	N701i		SH700i	902SH	P901iS	F901iC
	N700i		SH901iC	801SH	P900iV	F700i
	N900iG	**802N**	SH900i	**802SH**	P900i	RakuPhoneII
	N900i	V-N701				
	N900iS					
2G	N506iS	J-N51	*SH506iC*	V302SH	P506iC	F506i
	N506i	V601N		V501SH	P252iS	F505i(GPS)
	N253i			V603SH	P253i	F672i
					Lechiffon	
					Prosolid	

	SANYO			Toshiba	
3G/2.5G	DoCoMo	AU KDDI	Vodafone	AU/Tuka	Vodafone
	SA700iS	W32SA	V801SA	W31T	902T
		W31SA	J-SA701	W21T	
		W22SA			
2G	D253i	Sweets	V401SA	A5509T	V603T
	D506i	A5505SA	J-SA06	A5511T	V501T
		A5507SA	J-SA51	A5504T	V601T
			J-SA05		V303T
					V602T

	Sony Ericsson			Mitsubishi	
3G/2.5G	DoCoMo	Vodafone	AU KDDI	DoCoMo	Vodafone
		802SE	W32S	D701i	
			W31S	D901iS	
				D901i	
2G	Premini		A1404S/S	D253i	V301D
	Rediden		A1402S	D506i	
	SO506i				

Fig. 7.7 Categorization of handset models by vendor and operator (Sep. 2005)
Source: Website and catalogs of NEC, Sharp, Panasonic, Fujitsu, Mitsubishi, Sanyo, Toshiba, SonyEricsson [27–34]
*(italic) indicates "regional" models while others are "customized" models

Operator	Vender				Total		
		Customized	Global	Regional		High-end	Midrange
NTT DoCoMo	Fujitsu	9			9	5	4
	Mitsubishi	5			5	2	3
	NEC	7		1	8	5	3
	Panasonic	9			9	4	5
	Sharp	3		1	4	3	1
	Sony Ericsson	3			3		3
	Total	**36**		**2**	**38**	**19**	**19**
AU	Casio	1			1		1
	Kyocera	1			1		1
	Sanyo	3			3		3
	Sharp	3			3		3
	Sony Ericsson	10			10	8	2
	Toshiba	2			2		2
	Total	**20**			**20**	**8**	**12**
Vodafone	Mitsubishi	1			1		1
	NEC			1	1	1	
	Nokia		1		1		1
	Sanyo	1			1		1
	Sharp	5		1	6	2	4
	Sony Ericsson	1			1		1
	Toshiba	3			3		3
	Total	**11**	**1**	**2**	**14**	**3**	**11**
Total		67	1	4	72	30	42

Fig. 7.8 Service operators and handset vendors in Japan's mobile phone market by types and classes (Sep. 2005)

which model they should adopt is also a decision that will affect their share in mobile service market. For service operators, it is considered as an important strategy to obtain new subscribers and maintain the existing users by launching more attractive handset models [35]. Moreover, the models that the service operators adopt from the handset vendor to launch in the market also determine the structure of customers' subscription by digital segment.

Since the relationship between service operators and handset vendors is an important key factor to determine the provision of handset models in Japan's market, the ratio of high-end handset models adopted by the service operators also affect their own 3G subscription rate. As illustrated in Fig. 7.8, the ratios of high-end handset models by service operator are 50, 40 and 21% for NTT DoCoMo, au KDDI and Vodafone respectively. Consequently, by the end of September 2005, more than 70% of NTT DoCoMo subscribers communicated through 3G wireless technology and have the accessibility to various advanced application service while only 55% of au KDDI and 40% of Vodafone did so as illustrated in Fig. 7.9 [36].

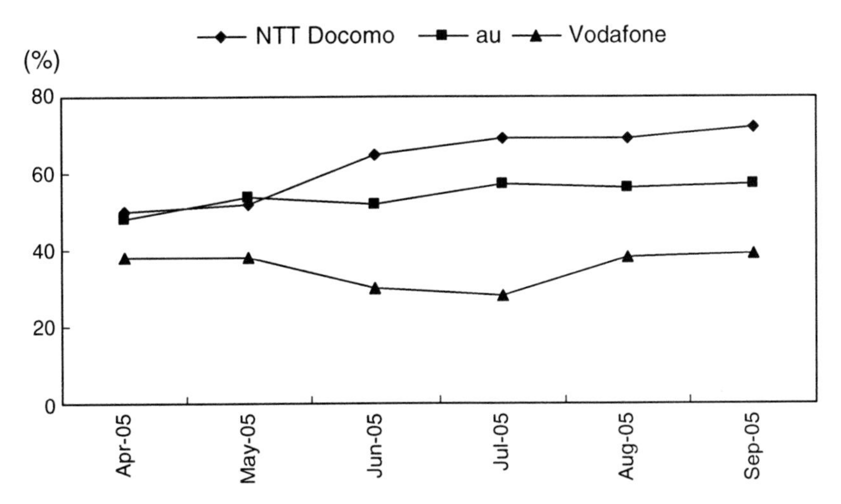

Fig. 7.9 Trend in the ratio of 3G mobile phone service subscribers of each service operator in Japan (Apr. 2005–Dec. 2005)
Source: Database of subscribers by service operator [36]

Since all models are sold with both the names of the service operator and the handset vendor, both the service operator and handset vendor bear responsibility for the handsets in Japan. In other countries where the consumers choose the handset models with only the handset vendor's name, the handset vendors can focus simply on satisfying the demand of the immediate customers. Contrary to such a structure, the first step of the handset vendor to keep a place in the market in Japan is to satisfy the service operators and persuade them to adopt the handset models. Consequently, each model must meet the technical need of the service provided by the service operator as well as the brand image of both the handset vendor itself and the service operator. One manager of Fujitsu confessed, "While we would like to cooperate with multiple service operators, too; it is really difficult to afford the huge amount of R&D invested separately for each service operator." The one-to-one close relationship between the handset vendors and service operators can be considered as a consequence of supply side co-evolution in order to maintain a high-quality and rapidly innovative process.

In order to satisfy the demand of picky consumers, service operators use a high standard in choosing handset vendors and are the "real" customers of handset vendors in Japan. Handset vendors must develop and produce handsets to meet the technical specification and infrastructure of the service operators, and they also have to design fashionable products that can attract the end-users. Consumers, handset vendors and service operators form a strongly intertwined relationship where handset vendors and service operators supervise each other and the consumers dominate both handset vendors and service operators as illustrated in Fig. 7.10. Customized production and model development combined the handset vendor and service operators tightly together leading their branding targets to converge in order to satisfy

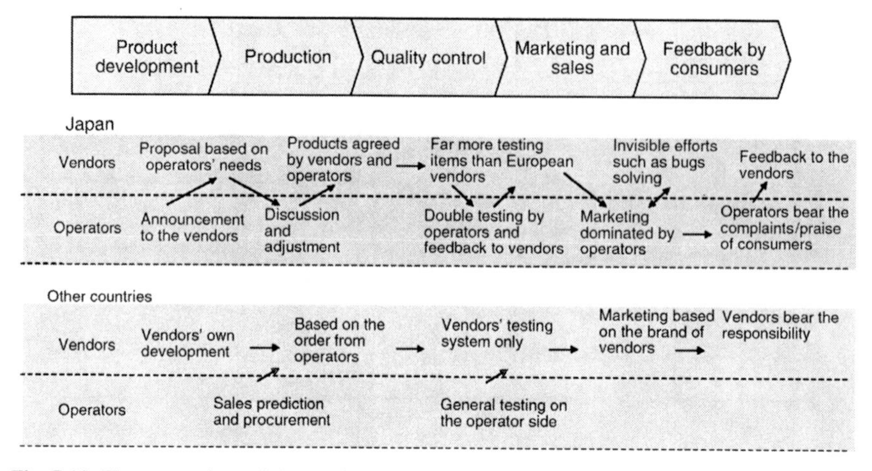

Fig. 7.10 The comparison of the mechanism under the mobile phone markets in Japan and other countries

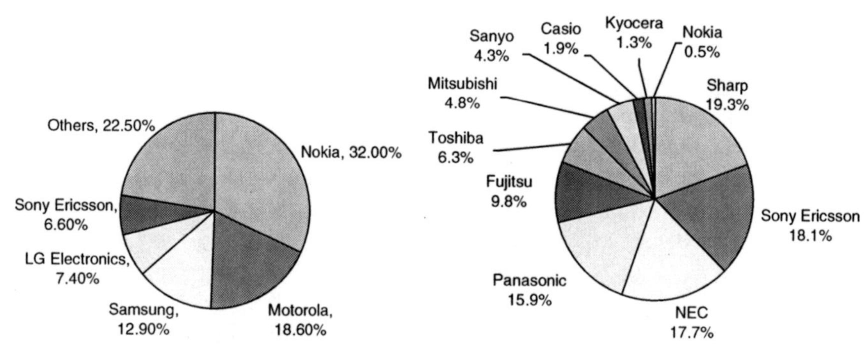

Fig. 7.11 Market share of handset in the worldwide market and Japan (3Q 2005)
Source: Press release of International Digital Corporation (IDC) and IDC Japan [37]

Japan's demanding customers' functionality requirements. This supply structure, together with customers' high level of demand, constitutes a sophisticated institutional system resulting in Japan's rapid increase in high functional mobile phones in a self-propagating way.

However, this structure, on the other hand, resulted in the development of a barrier against global handset vendors. By comparing the market share of 3Q 2005 in the worldwide and Japan's market, it is clear that top global handset vendors do not perform well in Japan's market as illustrated in Fig. 7.11. Since most global handset vendors provide global models in different countries, it is very difficult for them to become so intimately related to the service operators in Japan and enter the one-to-one order-made tight relation of Japan's mobile phone handset supply side.

7.4 A Virtuous Cycle Between Demand and Supply Sides Driving Japan's Co-Evolutionary Mobile-Driven Innovations and Institutions

Analyses in the preceding sections suggest that not only does the supply side provides the market with advanced hardware (features of handsets) and software (service such as mobile internet) but also the consumers play a significant role in inducing innovative product and services through their learning effects. Compared with other markets with high usage of ICT, Japan's customers appear to be more willing to regularly use a wider variety of features on their mobile phones. This aggressive customer behavior compels the handset vendors and service operators to provide handsets with high-quality as well as innovative features that, in turn, stimulate a demanding customers leading to a virtuous cycle between them. The extremely high penetration rate of the mobile phone camera illustrated in Fig. 7.3 and the high penetration rate of mobile e-mail shown in Fig. 7.12 can be considered the result of the virtuous cycle where consumers are willing to adopt new technology and handset vendors/service operators respond positively to demanding consumers.

Figure 7.12 demonstrates that while Japan's e-mail utilization ratio by means of mobile phones and PCs is similar at 87.7 and 94.2% respectively, in the US it is much lower for mobile phones at 12.4% while extremely higher for PCs at 96.1%. This contrast demonstrates the demanding nature of Japan's customers for

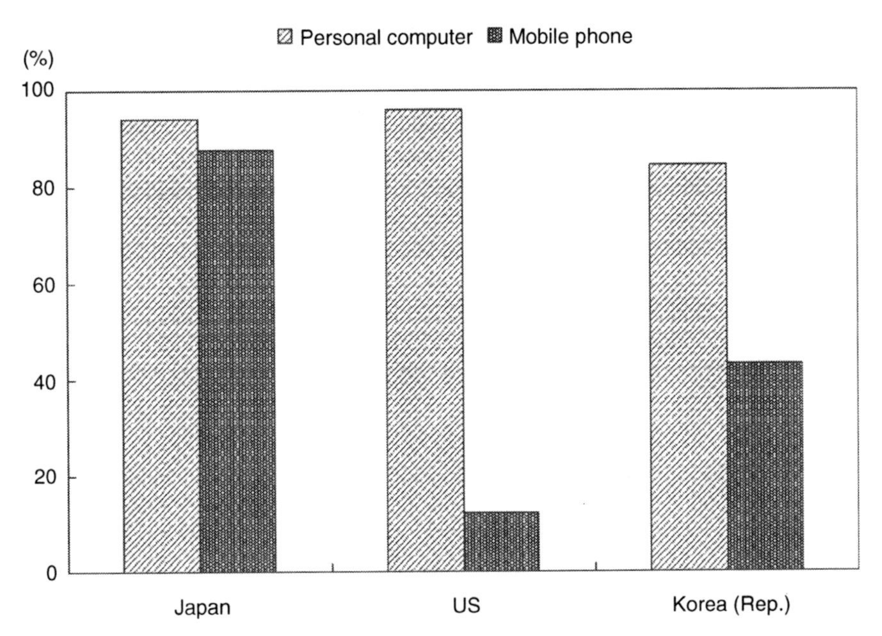

Fig. 7.12 Penetration rate of e-mail user by personal computer and mobile phone in Japan, US and Korea (Rep.) (2001–2005)
Source: 2005 White Paper of Information and Communications in Japan [2]

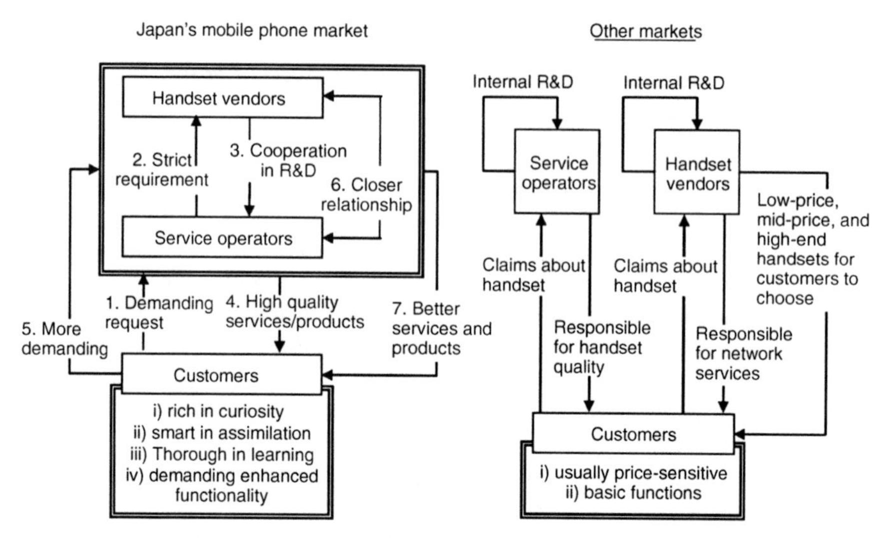

Fig. 7.13 Comparison of institutional systems in Japan's mobile phone market and in other markets

well-functioned mobile phones. The virtuous cycle constructed by both the demanding customer side and the closely intertwined supply side is illustrated in Fig. 7.13. The customers' demanding request urges service operators to set up stricter requirement toward handset vendors. Next, in order to satisfy the service operators, handset vendors have no choice but to conduct cooperation in R&D with service operators. Such close relationship enables the supply side to provide the customers with high quality services and products. However, customers with high learning capability get used to such high quality and become more demanding. Consequently, further R&D cooperation of the supply side and R&D investment tie the handset vendors and service operators more tightly and enables the supply of even better services and products. These processes construct the virtuous cycle in Japan's mobile phone market. On the other hand, the structure of other markets is different from that of Japan. Service operators are responsible only for telecommunication services, and handset vendors offer the handsets according to their own R&D development and marketing strategy. Without the close relationship with the service operators, various handset vendors offer various kinds of handsets, including low-price, mid-price and high-price handsets, directly to the customers to choose by their own concerns such as price factors.

On the basis of this virtuous cycle, Japan's noticeable advancement in mobile-driven innovation emerged and changed its institutional systems in a way to induce further innovation. Thus, the co-evolutionary dynamism between them can be emerged. A virtuous cycle between demand side and supply side plays a significant role in reactivating Japan's indigenous co-evolutionary dynamism between mobile-driven innovation and institutional change. In this marketplace, consumers play a distinctive role bringing to this cycle unique characteristics such as being (a) rich in curiosity, (b) clever in assimilation, (c) thorough in learning, and (d) demanding enhanced functionality play a leveraging role.

7.5 Conclusions

7.5.1 General Summary

This research elucidates the specific structure of the mobile phone market in Japan as follows. The first part of the analysis classified the existing handset models showing that high-end handsets occupy the largest share in Japan's market, and that the ratio is much higher than the worldwide average level. The second part of this analysis demonstrated that most domestic handset vendors offer customized models to satisfy the specific demands of each service operator. Consequently, global handset vendors have difficulty entering the marketplace by simply offering global models to the service operators. Similarly, because of the unique institutional structure in Japan, particularly the mutual dependence of the handset vendors and service operators on each other for their success, the Japanese vendors/service operators also struggle in other markets. Due to Japan's social institution, consumers' strong consciousness toward high quality, and innovative functions driving both service operators and handset vendors' commitment to quality, a closed but high-standard relationship between handset vendors and service operators has been developed which works closely with consumer demand. This mutually dependent relationship pushes the virtuous cycle of technological innovation so that it works smoothly and efficiently.

7.5.2 New Findings

The following are new findings of particular note based on this analysis:

(a) Japan has become a market dominated by high-end handsets since customers in Japan prefer high quality and new functionality. According to the learning curve, it is observed that the start of mobile phone camera is the most important time point that triggered the learning coefficient turned upward.
(b) Japan has also become a market dominated by order-made handset development between service operators and handset vendors whose close tie-up supports this demanding market. Both the service operators and handset vendors are responsible for the quality of handset, and the service operators, not the mobile phone end-users, are actually the immediate customers of the handset vendors.
(c) The interaction of the demand side and supply side in the mobile phone market of Japan formed a co-evolution mechanism. It enables the extraordinarily high achievements in terms of quality and level of technology in this unique marketplace. However, on the other hand, the marketplace is relatively closed and also prevents other global handset vendors from succeeding in Japan making it difficult for Japanese vendors/operators to succeed in other markets, either.

7.5.3 Policy Implications

Based on these unique findings, there are important implications for policy makers inside and outside Japan who are interested in creating an environment that stimulates an innovative marketplace of similar attributes. The following are some of the major policy implications:

(a) The first lesson learned from Sect. 7.2 is that one or two groundbreaking but user-friendly function expansion is necessary to stimulate the demand of a saturated market. The mobile camera feature combines handset and digital camera, and it showed the market a new possibility for using the mobile phone device. However, too many minor functions might only cause the consumers to spend more time to adapt and should be avoided.

(b) Handset vendors in Japan rely on the service vendors very much. This close relationship of R&D, manufacturing and sales cooperation drives the progress of the mobile phone in Japan, but may not be applicable in other markets where the service operators and handset vendors have to face the market separately. Japanese vendors and service operators should learn to be more independent from each other and more flexible in order to adapt themselves to the local supply side structure in other markets.

(c) Although Japanese handset vendors and service operators are good at satisfying the demanding consumers in Japan with high-quality products and services, this high-cost business would be impossible without the sufficient support of consumers. The positive co-evolution mechanism in Japan builds on the foundation of a unique social institution. To succeed in other markets, Japanese businesses have to adapt their strategies to the attributes of each market and learn to balance the cost and quality requirement. Another possible way is to wait or educate the consumers until they also request services and products at the same level as Japanese consumers.

7.5.4 Future Works

Through this research, sources of Japan's co-evolutionary dynamism between mobile phone driven innovation and its unique institutional systems have been identified. While this dynamism induces the dramatic advancement of mobile phones by enhancing functionality in a self-propagating manner, it incorporates structural constraints in embarking in a global market where more flexible options of both supply and demand structure are required.

Given that Japan's sustainable development depends on its own economic co-evolution with global sustainability, Japan's mobile phone driven innovation should be globally expansive. Thus, Japan's mobile phone industry should shift from a homogeneously integrated structure that works well, but only in Japan, to a heterogeneously structure which is able to adapt and thrive amidst other demand side factors. While Japan's current system appears to be homogenous, even in the relatively high-standard market in Japan, there are still differences among major players with respect to their strategic positioning, technology standards decisions and global strategies. While NTT DoCoMo's focus is on maintaining domestic supremacy in the consumer market with 3G capabilities by targeting "discerning customers" who want high quality service, au KDDI seems to be positioning itself for broader applications. How such differences in strategy and positioning may affect their achievement in penetrating into other markets is still a critical question worthy of further exploration. After certain markets reach a saturation point, how to accelerate co-evolution process not only in the domestic market but also in other markets will determine the long-term success of the firms involved. Future work should focus on this critical dimension.

References

1. I. Atsushi, Estimating demand for cellular phone services in Japan, Telecommunications Policy 29 (2005) 3–23
2. F.A. Bass, New product growth model for consumer durables, Management Science (1969) 215–227
3. Business Week (2000 Jan. 17), Feature Article of i-mode (The McGraw-Hill Companies Inc., New York 2000)
4. C. Chen and C. Watanabe, Diffusion, substitution and competition dynamism inside the ICT market: a case of Japan, Technological Forecasting and Social Change 73, No. 6 (2006) 731–759
5. C. Easingwood, V. Mahajan and E.A. Muller, Nonuniform influence innovation diffuison model of new product acceptance, Marketing Science 2, No. 3 (1983) 273–295
6. J. Funk, The future of the mobile phone internet: an analysis of technological trajectories and lead users in the Japanese market, Technology in Society 27 (2005) 69–83
7. GfK Marketing Services Japan Ltd. (GfKMSJ), Panel survey: profidence of mobile market in Japan (GfKMSJ, Tokyo, 2005)
8. International Telecommunication Union (ITU), Asia-Pacific Telecommunication Indicators 2002 (ITU, Geneva 2002)
9. International Telecommunication Union (ITU), Internet Reports 2004: The Portable Internet (ITU, Geneva, 2004)
10. K. Jonsson and K. Miyazaki, 3G Mobile Diffusion in Japan: Technology strategies of au KDDI and NTT DoCoMo and Technology Adoptive Users, Journal of the Japan Society for Management Information 13 (2004) 57–77
11. D.B. Jun, S.K. Kim, Y.S. Park, M.H. Park and A.R. Wilson, Forecasting telecommunication service subscribers in substitutive and competitive environments, International Journal of Forecasting 18 (2002) 561–581
12. D.B. Jun and Y.S. Park, A choice-based diffusion model for multiple generations of products, Technological Forecasting and Social Change (1999) 45–58
13. M. Kodama, Strategic community-based theory of firms: case study of NTT DoCoMo, Journal of High Technology Management Research 19 (2003) 307–330
14. R. Kondo and C. Watanabe, The virtuous cycle between institutional elasticity, IT advancement and sustainable growth: can Japan survive in an information society, Technology in Society 25 (2003) 319–335
15. E. Mansfield, Technical change and the rate of imitation, Econometrica 29 (1961) 741–766
16. Ministry of Internal Affairs and Communications (MIAC) Japan, 2005 White Paper of Information and Communications in Japan (MIAC, Tokyo 2005)

17. M. Nagamachi, Kansei engineering as a powerful consumer-oriented technology for product development, Applied Ergonomics 33 (2002) 289–294
18. NTT Mobile Communications Network, Inc., NTT DoCoMo's Vision towards 2010 (NTT Mobile Communications Network, Inc., Tokyo 1999)
19. A. Slawsby and A.M. Leibovitch, Worldwide Mobile Phone 2005–2008 Forecast by Feature Tier (A Feature-Rich Future. Framingham: International Data Corporation, IDC, 2005)
20. J.W. Woodlock and L.A. Fourt, Early prediction of market success for grocery products, Journal of Marketing 25 (1960) 31–38
21. M. Yamada, R. Furukaw and M. Ishihara, Classification pattern of new product diffusion, Marketing Science by Japan Institute of Marketing Science 4 (1995) 16–36 [*in Japanese*]
22. Website of NTT DoCoMo, Product Information (September 2005) <http://www.nttdocomo.co.jp/>
23. Website of AU, Product Information (September 2005) <http://www.au.kddi.com/>
24. Website of Vodafone, Product Information (September 2005) <http://www.vodafone.jp/top.htm>
25. Website Telecommunications Carriers Association (TCA) Japan. <http://www.tca.or.jp/japan/database/daisu/index.html>
26. Website of Bank of Japan. Monthly Report of Price Index; 1996–2005 <http://www.boj.or.jp/stat/stat_f.htm>
27. Website of International Data Corporation (IDC, 2005) <http://www.idc.com>
28. Japan Information Processing Development Corporation (JIPDC), IT White Book (JIPDC, Tokyo 2003)
29. Website of NEC, Product Information (September 2005) <http://www.n-keitai.com/>
30. Website of Sharp, Product Information (September 2005) <http://www.sharp.co.jp/products/index.html>
31. Website of Panasonic, Product Information (September 2005) <http://panasonic.jp/mobile/>
32. Website of Fujitsu, "Product Information" (September 2005) <http://www.fmworld.net/>
33. Website of Mitsubishi, Product Information (September 2005) <http://www.mitsubishielectric.co.jp/products/index.html>
34. Website of Sanyo, Product Information (September 2005) <http://www.e-life-sanyo.com/>
35. Website of Toshiba, Product Information (September 2005) <http://www.toshiba.co.jp/digital>
36. Website of SonyEricsson, Product Information (September 2005) <http://www.sonyericsson.co.jp/>
37. Website of International Data Corporation (IDC) Japan; September 2005 <http://www.idcjapan.co.jp/Press/index.html>

Chapter 8
Technopreneurial Trajectory Leading to Bipolarization of Entrepreneurial Contour in Japan's Leading Firms

Abstract Contrary to the institutional non-elasticity against a new paradigm in the information society of the 1990s, the dramatic advancement of mobile technology has led Japan's technological development to surge to a new level. This has, in turn, changed the institutional system and subsequently, entrepreneurial contour.

This chapter attempts to elucidate the co-evolutionary dynamism between transformation of the characterization of technology and subsequent change in entrepreneurial contour in leading high-technology firms.

Empirical analyses are conducted based on the following:

(a) Co-evolutionary dynamism between the advancement of mobile phones with Internet access service and institutional change, and
(b) The transformation in entrepreneurial contour in leading Japanese and US firms facing the transition to a post-information society.

On the basis of the above analyses, it is demonstrated that the characterization of technology has transformed as a consequence of the paradigm shift from an industrial society to an information society in the 1990s, which in turn has reshaped entrepreneurial contour in leading high-technology firms. This suggests that consistent market learning efforts and innovation generation leading to an increase in technology elasticity to the price of technology is essential for a technopreneurial trajectory toward a post-information society.

8.1 Introduction

In contrast to the high-technology miracle of the 1980s, Japan's international competitiveness declined dramatically (IMD) [1]. This can be attributed to an incorrect selection of a development trajectory clinging to economic growth development

C. Watanabe, *Managing Innovation in Japan: The Role Institutions Play in Helping or Hindering how Companies Develop Technology*, DOI: 10.1007/978-3-540-89272-4_8,

model due to organizational inertia in industrial society development and a subsequent systems conflict in a paradigm change to an information society [2]. However, contrary to such institutional non-elasticity against a new paradigm in the information society of the 1990s, the dramatic advancement of mobile technology has led Japan's technological development to surge to a new level [3–5]. This has, in turn, changed the institutional system and subsequently entrepreneurial features [6–8].

To date, not a few studies analyzed the institutional sources governing nation's competitiveness [9, 10]. In addition, a number of studies have identified the impacts of the advancement of IT on nation's economic competitiveness [8, 11, 12]. Watanabe et al. (2004) [13, 14] pointed out the substantial differences of the characterization between manufacturing technology and IT. They also pointed out that the systems conflict that Japan experienced in a paradigm change from an industrial society to an information society can be attributed to such differences and subsequent disengagement between the advancement of IT and its traditional institutional systems. While their analysis provides a reasonable answer to the reactivation of Japan's co-evolutionary dynamism in mobile phone driven innovation, they have not taken the profound effects that the transformation of the characterization of technology might provide on entrepreneurial features.

> This paper, prompted by the notable contrast observed in (a) mobile phone diffusion in contrast to that of fixed phone [6], (b) a conspicuous growth in functional fine ceramics in contrast to a stagnation in structural fine ceramics [15], and (c) bipolarization trajectory in leading high-technology firms [7] attempts to elucidate the co-evolutionary dynamism between transformation of the characterization of technology and subsequent change in entrepreneurial features.

Empirical analyses are conducted based on the following:

(a) Co-evolutionary dynamism between the advancement of mobile phones with Internet access service and institutional change,
(b) Elucidation of the black box of new functionality development and its mechanism by means of cross-product technology spillover dynamism in high-performance fine ceramics, and
(c) The transformation to technopreneurship in leading Japanese firms confronting the transition to a post-information society.

Section 8.2 demonstrates a co-evolutionary dynamism between mobile phone driven innovation and institutional systems. Section 8.3 reveals transformation in entrepreneurial contour in Japan's leading firms facing the transition to a post information society. Section 8.4 briefly summarizes new findings and policy implications.

8.2 Co-Evolutionary Dynamism Between Mobile Phone Driven Innovation and Institutional Systems: *Cumulative Learning Leading to Creating New Functionality*

8.2.1 Dramatic Decline in Japan's Productivity in an Information Society in the 1990s

In contrast to the high-technology miracle of the 1980s, Japan's technology productivity declined dramatically in the 1990s which, despite being the world's highest R&D intensity, resulted in a dramatic decrease in TFP (total factor productivity) as illustrated in Fig. 8.1.

A dramatic decline in marginal productivity of technology (MPT) can be attributed to an incorrect selection of a development trajectory and a subsequent systems conflict in a paradigm change to an information society as illustrated in Fig. 8.2.

8.2.2 Mobile Phone Driven Innovation Emerged in the 2000s

However, a surge in co-evolutionary dynamism has been evident in recent years due to mobile phone driven innovation. Mobile phone subscriptions exceeded those of fixed phones in 1998 in Japan after a dramatic increase in market growth from 1991 as illustrated in Fig. 8.3.

Triggered by i-mode service introduced in February 1999, IP mobile phone diffusion was so rapid that the number of IP mobile phones in use exceeded that of non-IP mobile phones at the beginning of 2001 as illustrated in Fig. 8.4.

Thus, dependency on IP mobile phones has accelerated and resulted in a utilization rate of 94.1% relative to all mobile phones, which is the world's highest rate and is conspicuously high compared to the 33.5% level in the US as illustrated in Fig. 8.5.

Advancement of IT characterized by mobile phone substitution for fixed phones and increased Internet dependency has led to a dramatic change in Japan's lifestyle pattern. In the last 2 years, communication between family members and friends has changed from face-to-face conversation to e-mail as illustrated in Fig. 8.6.

Such intensive interaction with institutions increased the learning coefficient from the middle of 2002 as illustrated in Fig. 8.7.

An increased learning coefficient enhances new functionality development [16], leading to an increase in MPT and subsequent TFP growth as illustrated in Fig. 8.8.

As a consequence of cumulative learning based on intensive interaction with institutions, co-evolutionary dynamism between high dependency on IP mobile phones and new functionality emerged in line with the self-propagating dynamism of IT as illustrated in Fig. 8.9.

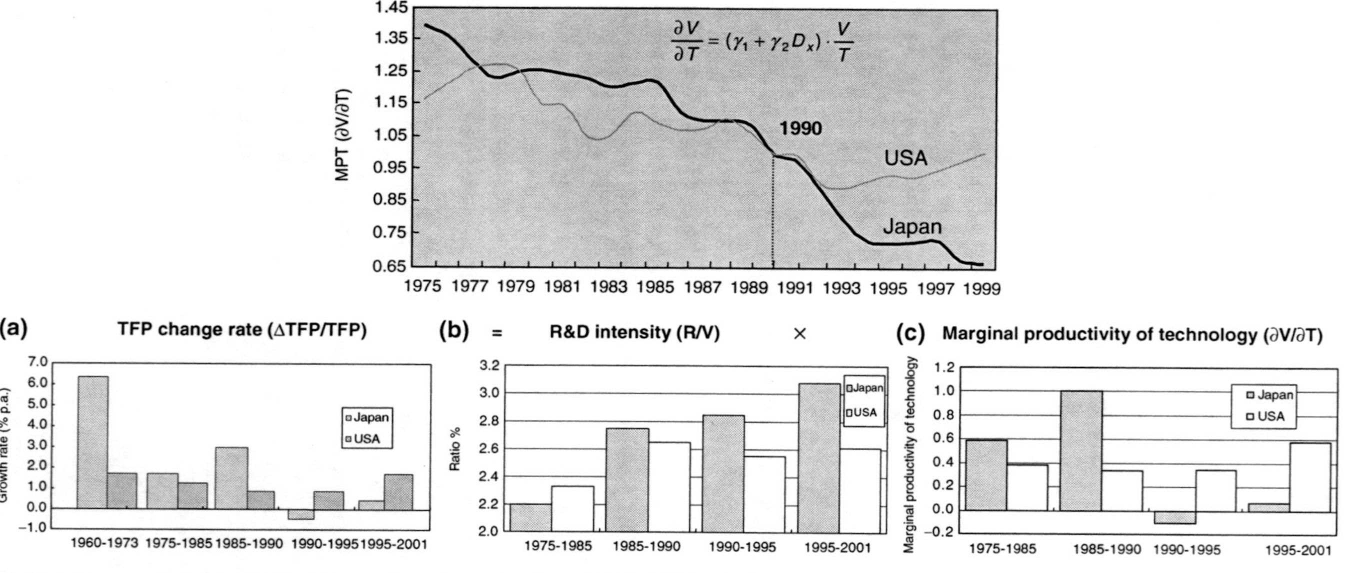

Fig. 8.1 Trend in marginal productivity of manufacturing technology (1975–1999) – index: 1990 = 1. (**a**) Trends in TFP growth rate in Japan and US (1960–2001). (**b**) Trends in R&D intensity in Japan and US (1975–2001). (**c**) Trends in marginal productivity of technology in Japan and US (1960–2001)

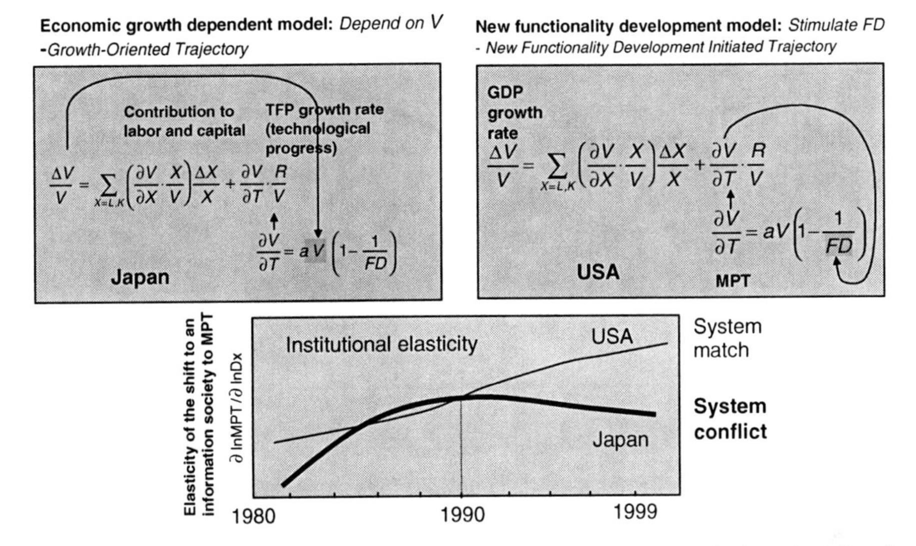

Fig. 8.2 Trends in elasticity of the shift to an information society to marginal productivity of technology (1980–1999) – Index: 1990 = 100

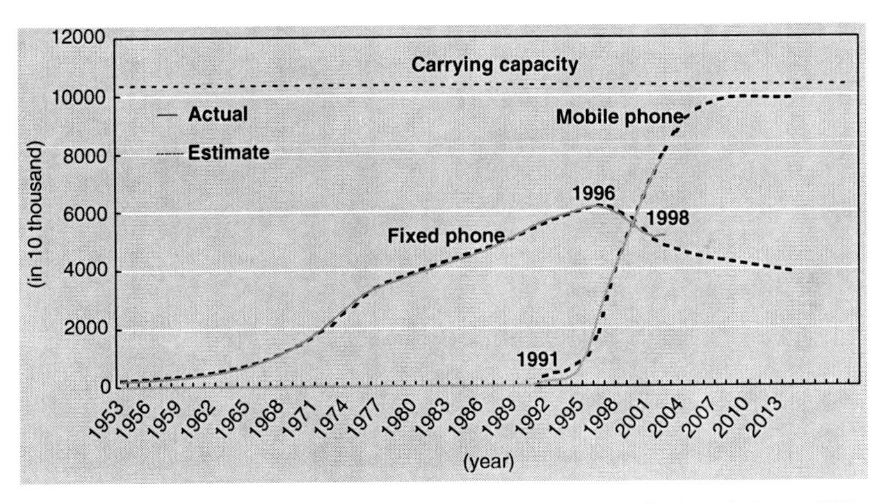

Fig. 8.3 Trend in the substitution trajectory of mobile subscribers for fixed line in Japan (1953–2004: actual, and 2005–2015: estimate)

In this dynamism, stimulated by severe competition among vendors and operators motivated by xenophobia in the process of diffusion, Japan's mobile phones have successively incorporated new functions as illustrated in Fig. 8.10.

Mobile phones with a camera function have increased over the last 4 years from 2.2 million subscribers (3.3%) in 2001 to 66.4 million subscribers (76.3%) in 2005 as illustrated in Fig. 8.11. Such innovation, in turn, has changed lifestyle patterns, daily customs, study and shopping habits as well as business models.

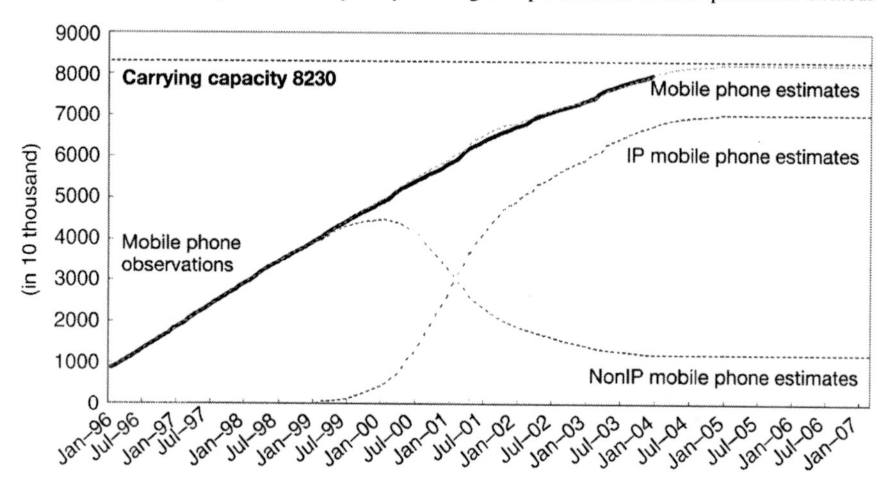

Fig. 8.4 Trend in mobile phone diffusion in Japan (Jan. 1996–Dec. 2003: actual, and Jan. 2004–Jan. 2007: estimate)

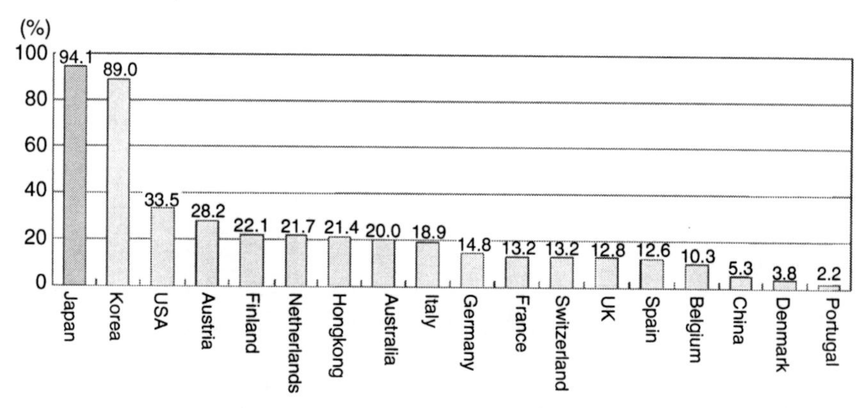

Fig. 8.5 Ratio with IP mobile phone in leading countries (end of Sep. 2004)

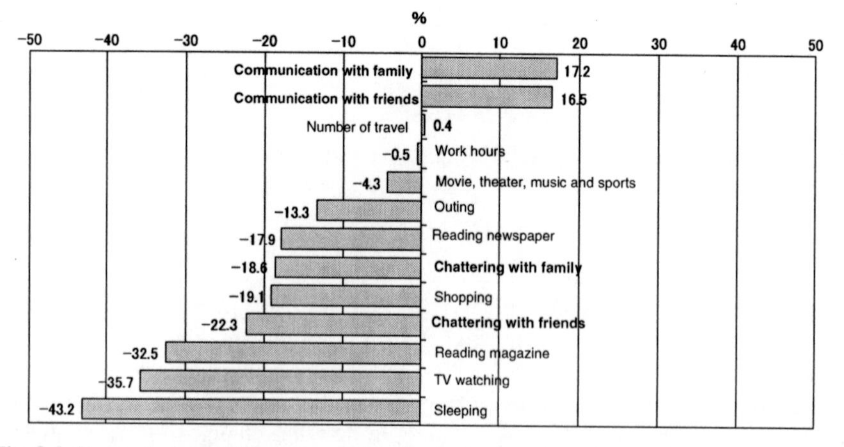

Fig. 8.6 Change in Japan's life pattern driven by the internet dependency (change between 2004 and 2002)

[a] Balance between the % of increase and decrease

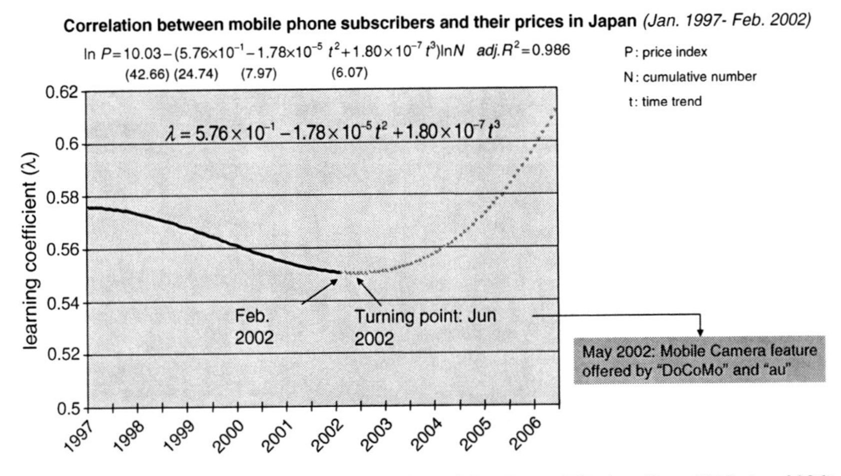

Fig. 8.7 Trend in learning coefficient in Japan's mobile phone diffusion (Jan. 1997–Jun 2006) – estimate (Jan. 1997–Feb. 2002) and extended estimate (Mar. 2002–Jun 2006)

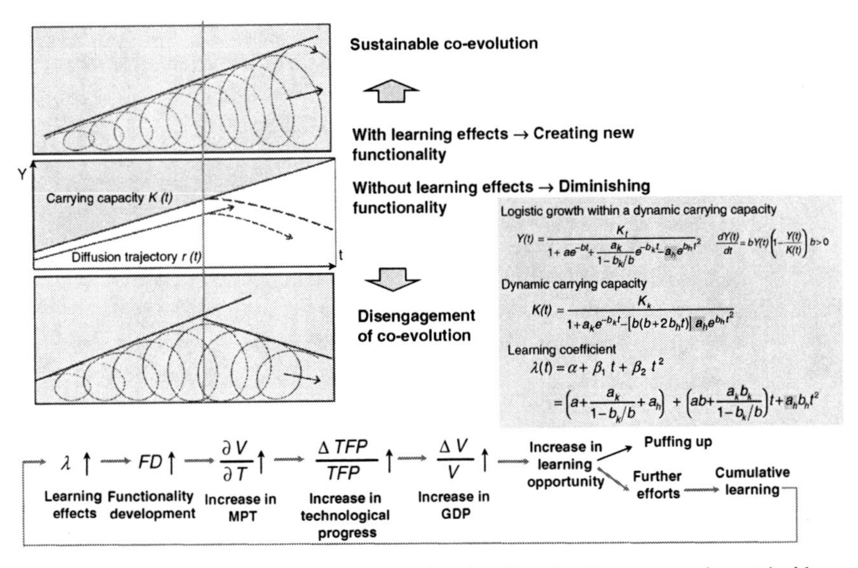

Fig. 8.8 Dynamism between learning, new functionality development, and sustainable co-evolution

While Japan's e-mail utilization ratio by means of mobile phones and PCs is similar at 87.7 and 94.2% respectively, in the US it is much lower for mobile telephones at 12.4% while extremely higher for PCs at 96.1% as illustrated in Fig. 8.12. This can be attributed to differences in the institutional systems of the two nations.

Japan's institutional systems leading to a noticeable contrast in high-level mobile phone dependency can largely be attributed to their unique functions as (a) rich in curiosity, (b) smart in assimilation, (c) thorough in learning, and (d) demanding enhanced functionality.

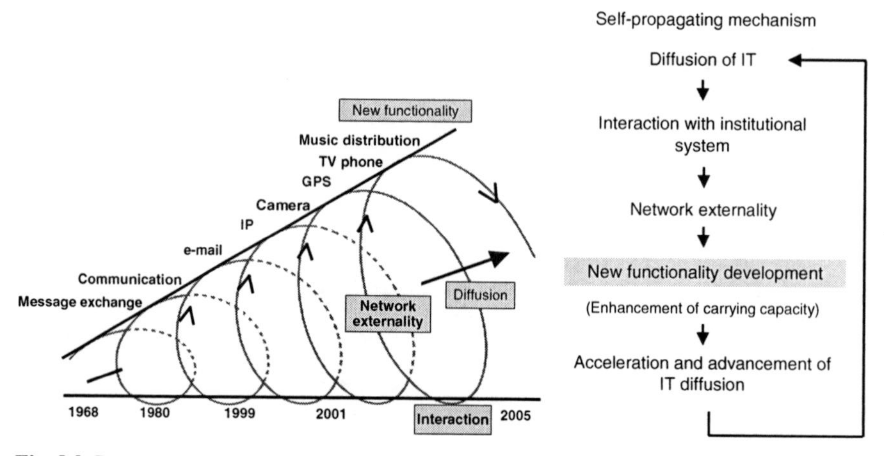

Fig. 8.9 Dynamism in creating IT's new functionality: interactive self-propagating mechanism

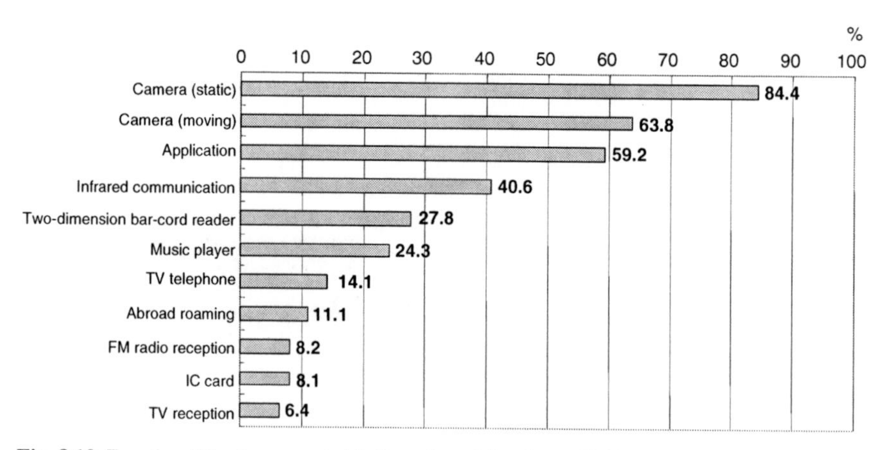

Fig. 8.10 Functionalities incorporated in Japan's mobile phone (2004)

Thus, co-evolutionary dynamism between mobile phone driven innovation and institutional change emerged in Japan at the beginning of the 2000s as illustrated in Fig. 8.13.

8.2.3 Swell of Reactivation of the Co-Evolutionary Dynamism

As a consequence of mobile phone driven co-evolutionary dynamism, a swell of reactivation of the co-evolutional function indigenously incorporated in Japan's way of MOT (Management of Technology) can be observed as illustrated in Fig. 8.14.

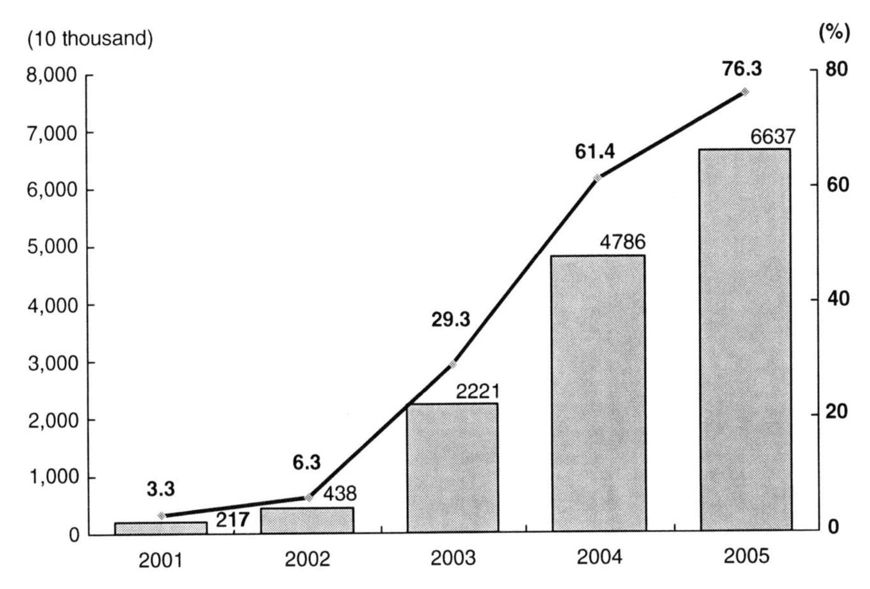

Fig. 8.11 Trend in mobile phones with camera function in Japan (2001–2005)
Source: White Paper 2005 on Information and Communication in Japan (MIC, 2005) [17]

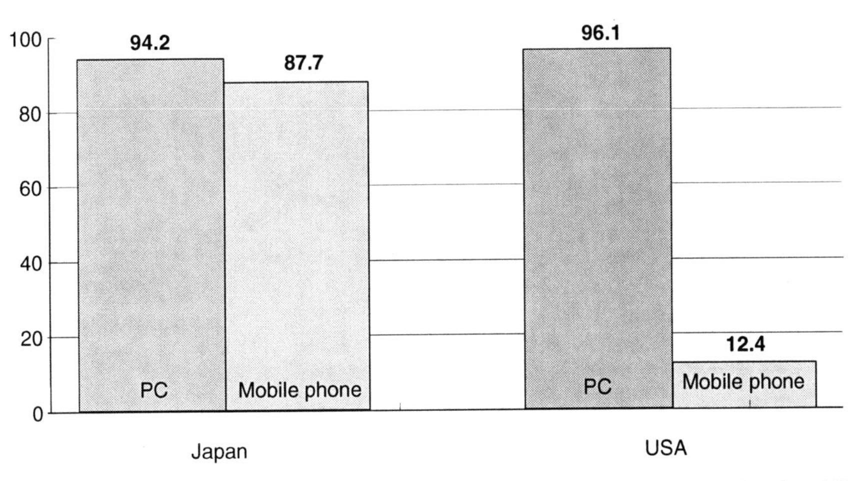

Fig. 8.12 Comparison of e-mail utilization ratio between Japan and the US by PC and mobile phone (2004)

Supported by such a reactivation, Japan has had a higher growth rate in manufacturing production and firm R&D investment than the US in a post-information society from 2001 as compared in Fig. 8.15 and Table 8.1.

The state of financial management of firms represented by the ratio of capital stock and gross stock has also reversed. While US firms maintained higher growth up to the fourth quarter of 1997, Japanese firms took the lead from the first quarter of 1998 as compared in Fig. 8.16 and Table 8.2.

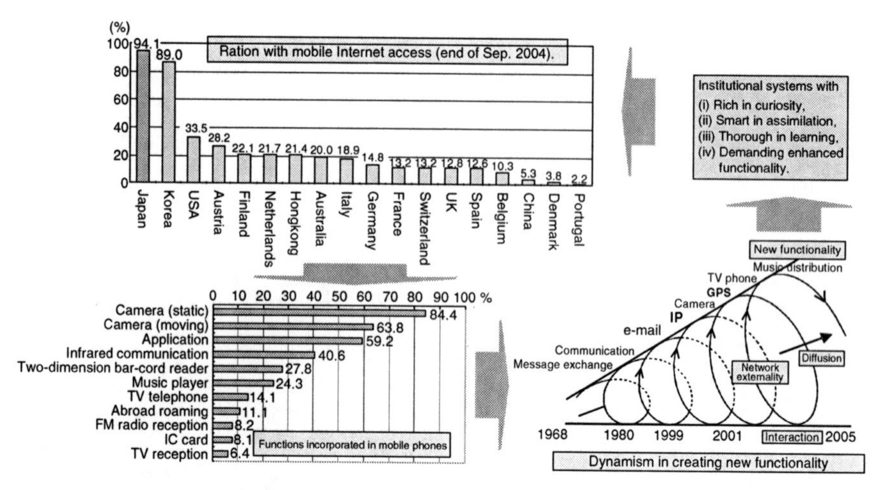

Fig. 8.13 Co-evolution of mobile phone driven innovation

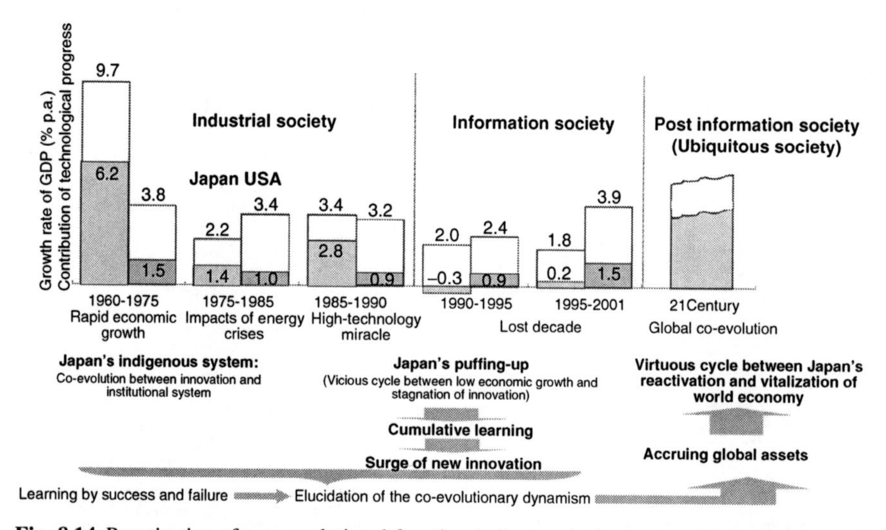

Fig. 8.14 Reactivation of a co-evolutional function indigenously incorporated in Japan's way of management of technology

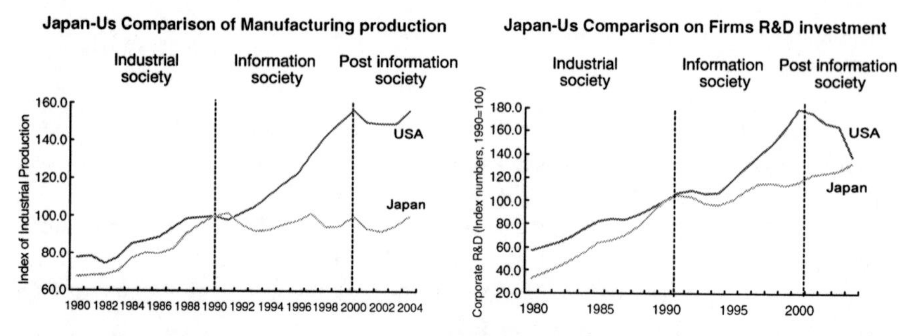

Fig. 8.15 Trend in manufacturing production and firms R&D investment between Japan and USA (1980–2004) – Index: 1990 = 100

Table 8.1 Balance of the growth rate of manufacturing production and firms relative R&D investment level between USA and Japan over the three periods (1980–2004): US growth rate – Japan's growth rate

1980–1990 (%p.a.)	1991–2000 (%p.a.)	2001–2004 (%p.a.)
−1.1	5.3	−1.0
−5.2	4.2	−9.8

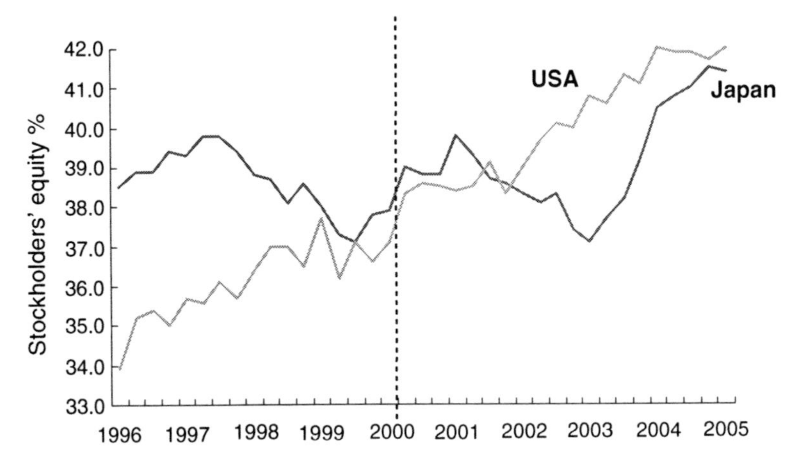

Fig. 8.16 Trend in Manf. Ind. Capital Stock Ratio between Japan and USA (first quarter 1996–first quarter 2005)

Table 8.2 Balance of the growth rate of manufacturing industry's capital stock ratio between US and Japan (1980–2004)

Information society		Post inf. society
Q1 1996–Q4 1997	*Q1 1998–Q2 2001*	*Q3 2001–Q1 2005*
0.53%	−0.28%	−0.31%
(2.1%p.a.)	(−1.1%p.a.)	(−1.3%p.a.)

Such a reversal can be typically observed in automotive industry net income as illustrated in Fig. 8.17. While Japan's leading three firms net income was behind the level of US and European leading three firms in 1999, Japan's three firms exceeded other three firms in 2004.

8.2.4 Implications

While Japan maintained the world's highest level of R&D intensity, due to institutional inelasticity to a new paradigm in an information society, its marginal productivity of technology declined dramatically in the 1990s, resulting in a dramatic

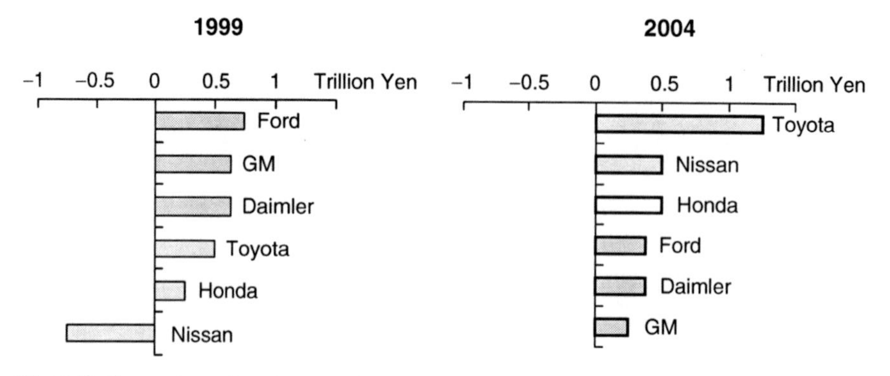

Fig. 8.17 Comparison of net income of world leading six automotive firms in 1999 and 2004

decrease in its TFP growth rate. However, the advancement of mobile phone driven innovation corresponded to the self-propagating nature of IT, leading to the construction of co-evolutionary dynamism between mobile phone driven innovation and institutional change. This can largely be attributed to intensive cumulative learning efforts with the following unique functions:

(a) Motivated by xenophobia,
(b) Rich in curiosity, smart in assimilation, thorough in learning and demanding enhanced functionality, and
(c) Stimulating effective assimilation of spillover knowledge leading to creating new functionality.

8.3 Transformation in Entrepreneurial Contour in Leading Firms Facing the Transition to a Post Information Society: *Output-Oriented R&D Based on External Acquisition*

8.3.1 Co-Evolution in High-Technology Firm Technopreneurial Structure

In leading high-technology firms, a co-evolutionary structure between technopreneurial dynamism and market value can be realized as illustrated in Fig. 8.18.

As a consequence of the market competition, Japan's electrical machinery firms share the similar R&D intensity as illustrated in Fig. 8.19. Figure 8.19 demonstrates that coefficient of 0.048 is equivalent to R&D intensity which is similar to Japan's whole electrical machinery R&D intensity in 2003, 4.4%.

While Japan's high-technology firm technopreneurial positions converged in an industrial society, they have shifted divergently in an information society, resulting in a bi-polarization structure in a post-information society as illustrated in Fig. 8.20.

Technological progress $W = F(X, Y)$ X: ratio of R&D to OI; Y: OI to sales.

Taylor expansion $\ln W = a + b \ln X + c \ln Y + d \ln X \cdot \ln Y$
$\ln W, \ln X, \ln Y \rightarrow \Delta TFP/TFP, R/OI, OI/S$.

$$\frac{\Delta TFP}{TFP} = a + b\frac{R}{OI} + c\frac{OI}{S} + d\frac{R}{OI}\cdot\frac{OI}{S} = a + b\frac{R}{OI} + c\frac{OI}{S} + d\frac{R}{S} = a + \frac{b}{OI/R} + c\frac{OI}{R}\cdot\frac{R}{S} + d\frac{R}{S}$$

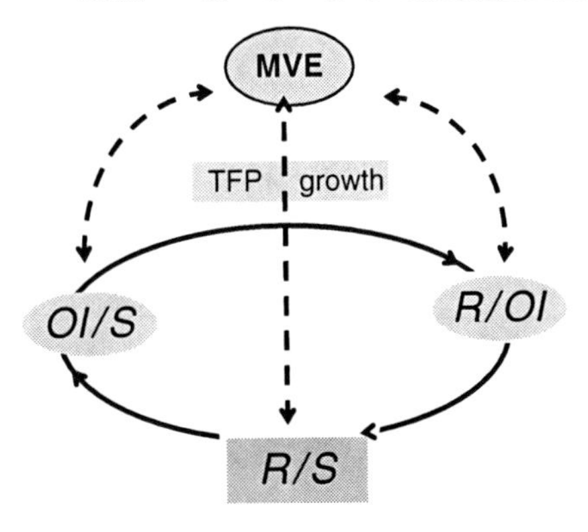

1. OI/S reflects MVE.
2. Higher MVE enables securing R from market.
3. This leads to higher R/OI.
4. Which results in higher R/S leading to OI/S increase.
5. All contributes to TFP growth which is sensitive to

(i) Subtle balance between OIS and ROI,
(ii) Substitute/complement relation between R/S and OI/R.

Fig. 8.18 Co-evolutionary structure between technopreneur dynamism and market value of equity

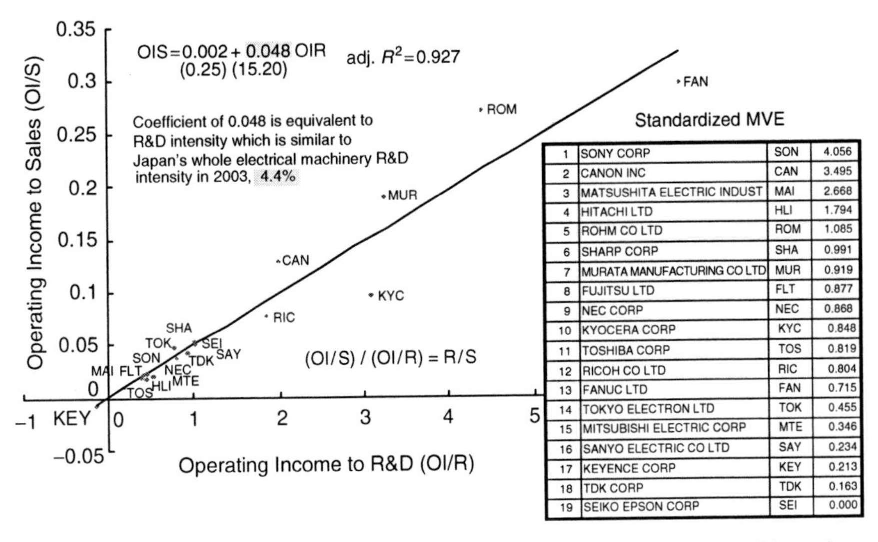

1	SONY CORP	SON	4.056
2	CANON INC	CAN	3.495
3	MATSUSHITA ELECTRIC INDUST	MAI	2.668
4	HITACHI LTD	HLI	1.794
5	ROHM CO LTD	ROM	1.085
6	SHARP CORP	SHA	0.991
7	MURATA MANUFACTURING CO LTD	MUR	0.919
8	FUJITSU LTD	FLT	0.877
9	NEC CORP	NEC	0.868
10	KYOCERA CORP	KYC	0.848
11	TOSHIBA CORP	TOS	0.819
12	RICOH CO LTD	RIC	0.804
13	FANUC LTD	FAN	0.715
14	TOKYO ELECTRON LTD	TOK	0.455
15	MITSUBISHI ELECTRIC CORP	MTE	0.346
16	SANYO ELECTRIC CO LTD	SAY	0.234
17	KEYENCE CORP	KEY	0.213
18	TDK CORP	TDK	0.163
19	SEIKO EPSON CORP	SEI	0.000

Fig. 8.19 Correlation between OIR and OIS in Japan's leading 19 electrical machinery firms (2001–2004 average)

8.3.2 Bi-Polarization in Technopreneurial Situation

With these notable observations, utilizing a co-evolutionary dynamism between OIS, ROI, RS, and TFP, a technopreneurial trajectory function can be developed as depicted in Fig. 8.21.

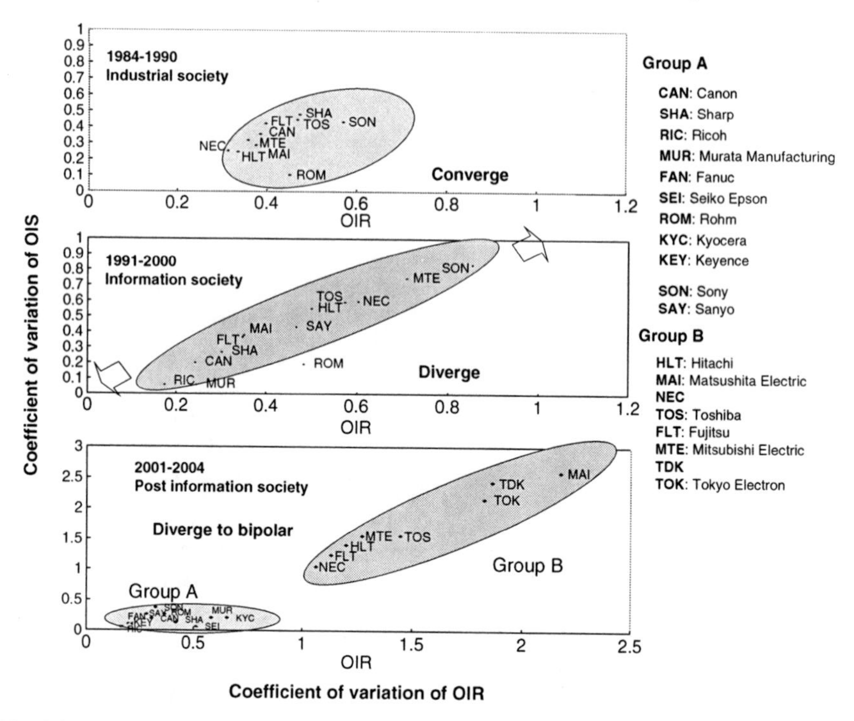

Fig. 8.20 Trends in coefficient of variations of OIR and OIS

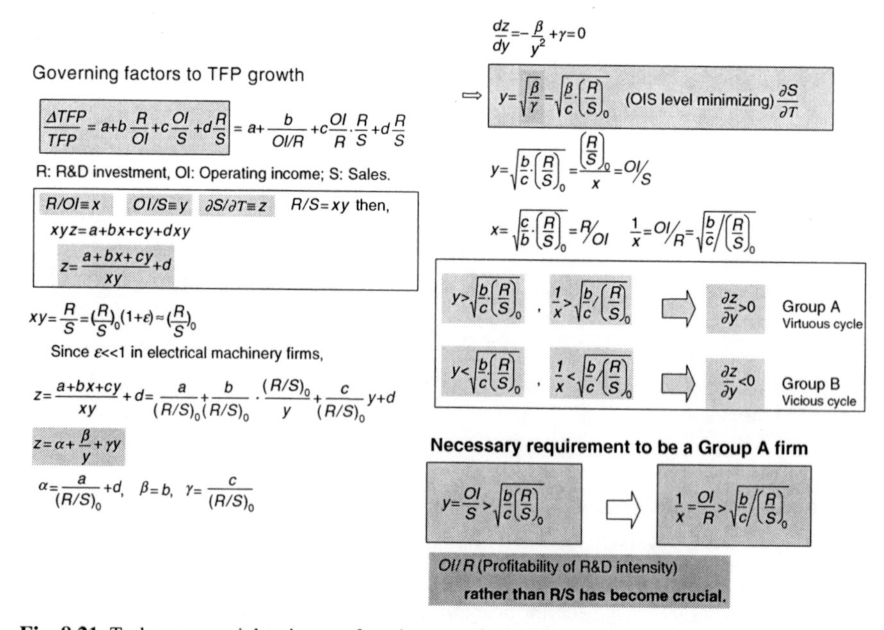

Fig. 8.21 Technopreneurial trajectory function – analytical framework

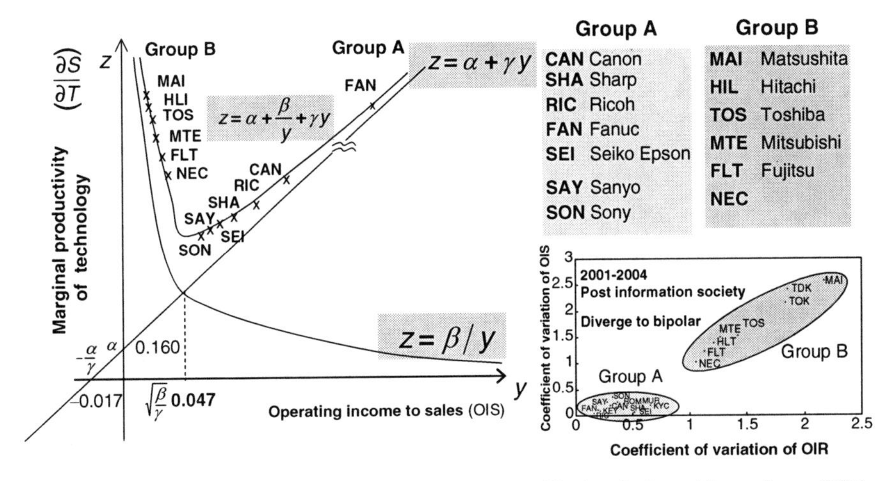

Fig. 8.22 Techno-preneurial situation of Japan's leading 13 electrical machinery firms (2001–2004)

A technopreneurial trajectory function demonstrates a bi-polarization structure of high-technology firms classified into Group A and Group B. While Group A firms enjoy a virtuous cycle between OIS and MPT (marginal productivity of technology) by means of OIR substitution for RS supported by intensive external learning efforts, Group B firms experience a vicious cycle as they cling to complement between OIR and RS as demonstrated in Fig. 8.22.

8.3.3 External Acquisition for OIR Substitution for R&D Intensity

A contrast in OIR substitution for RS in Group A firms and complement in Group B firms can be clearly demonstrated as compared in Fig. 8.23.

Given the same sales, OIR substitution for RS implies simply transferring a certain portion of sales, A from R to OI. Generation of higher OI (OI + A) by a smaller R (R – A) can only be enabled by acquiring external technology (A + α) as illustrated in Fig. 8.24.

Figure 8.25 and Table 8.3 demonstrate the contrast of external acquisition by means of market learning between firms in Group A (as Canon and Sharp) and B (as Hitachi and Matsushita). Cumulative market learning efforts have recently born fruit in improving their learning coefficient as demonstrated in Fig. 8.25.

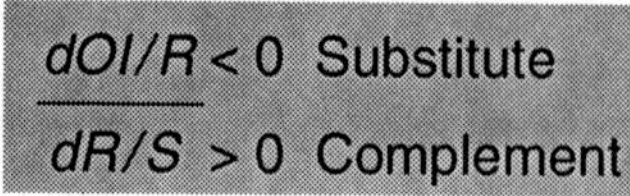

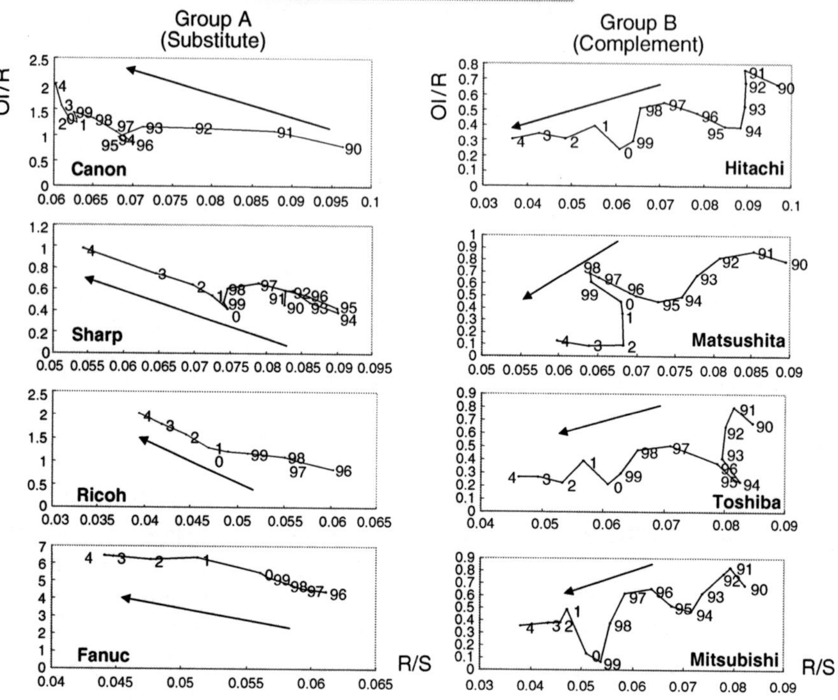

Fig. 8.23 Correlation between R&D intensity and operating income to R&D in Japan's eight electrical machinery firms (1990–2004) – 3 years moving average

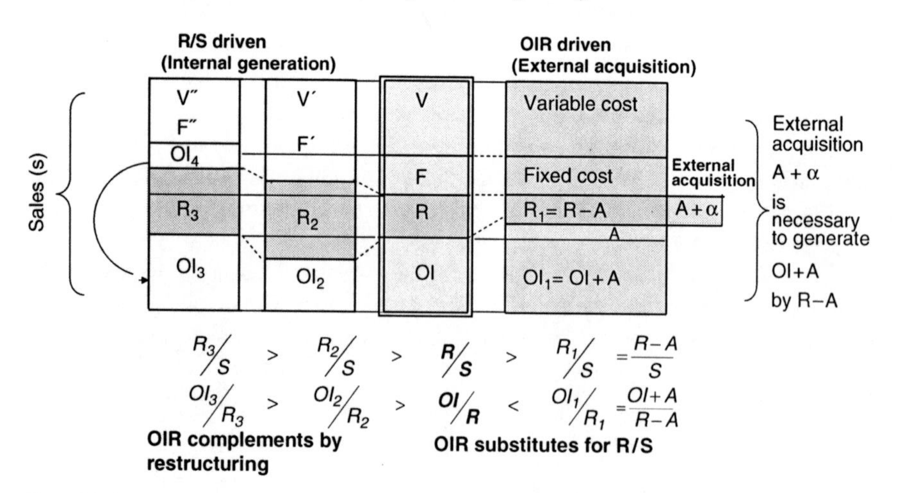

$$\frac{R_3}{S} > \frac{R_2}{S} > \frac{R}{S} > \frac{R_1}{S} = \frac{R-A}{S}$$

$$\frac{OI_3}{R_3} > \frac{OI_2}{R_2} > \frac{OI}{R} < \frac{OI_1}{R_1} = \frac{OI+A}{R-A}$$

Fig. 8.24 Scheme of OIR substitution for R/S by external acquisition

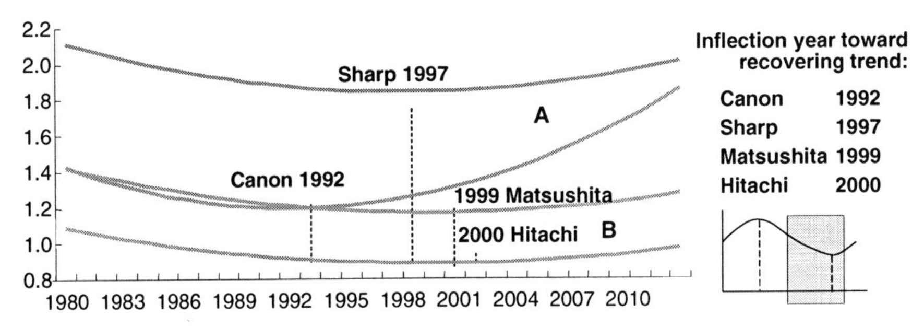

Fig. 8.25 Learning coefficients in four electrical machinery firms (1980–2003)

Table 8.3 OIR substitution for R/S and learning in Japan's four electrical machinery firms

		Substitution coefficient (b)[a]	Learning 2004/1990	Coefficient inflection year	Internal experience	External acquisition
A	Canon	−24.51	1.25	1992	Technology diversification	Market learning
	Sharp	−13.98	1.01	1997		
B	Hitachi	6.42	0.94	2000	NIH[b] syndrome	
	Matsushita	14.58	0.97	1999		

[a]Indicates coefficient b in the equation $\mathrm{OIR} = a + b\,R/S$. $\frac{dOI/R}{dR/S}$ < 0 substitute, > 0 complement
[b]Not invented here

8.3.4 Implications

All forgoing analyses demonstrate the bi-polarization of Japan's leading electrical machinery firms confronting the new paradigm to a post-information society leading to a transformation in entrepreneurial contour, and suggest new technopreneurial strategies.

Confronting the new paradigm to a post-information society, new technopreneurial strategies have been highlighted that lead to a transformation in entrepreneurial contour as follows:

(a) OIR substitution for RS,
(b) By means of shifting from internal generation oriented to effective integration of indigenous innovation and external acquisition market learning efforts.
(c) That changes entrepreneurial features from homogeneous to non-homogeneous features,
(d) Thus, an institutional shift from co-existence or co-adaptation of homogeneous firms to co-evolution between heterogeneous firms.

8.4 Conclusion

In light of a swell of the reactivation of Japan's co-evolutional dynamism leading to a noticeable contrast in high-level mobile phone dependency and subsequent mobile phone driven innovation, this paper attempted to elucidate this co-evolutionary dynamism.

Prompted by the notable contrast observed in a rapid increase in mobile phone diffusion, a conspicuous growth in high-functional fine ceramics, and also bi-polarization trajectory in leading high-technology firms, an elucidation of the co-evolutionary dynamism between transformation of the characterization of technology and subsequent change in entrepreneurial features was focused leading to the following empirical analyses:

(a) Co-evolutionary dynamism between the advancement of mobile phones with internet access service and institutional change,
(b) Elucidation of the black box of new functionality development and its mechanism by means of cross-product technology spillover dynamism in high-performance fine ceramics, and
(c) The transformation to technopreneurship in leading Japanese firms confronting the transition to a post-information society.

On the basis of the above analyses, noteworthy findings were obtained including:

(a) Transformation of characterization of technology through the course of interaction with institutions should be taken into firms technopreneurial strategies,
(b) Institutional elasticity (flexibility, adaptability, cooperative and openness to foreign ideas) is essential to correspond to such strategies,
(c) Mobile phone driven innovation induces a co-evolutionary dynamism between innovation and institutions leading to leverage the reactivation of Japan indigenous co-evolutionary dynamism,
(d) Cross functional spillover emerges mutation which can be enabled by spillover dynamism initiated by interaction of researchers,
(e) Cumulative learning by interactions would be the key in incorporating the new functionality as it constructs the following virtuous cycle:
Cumulative learning → Functionality development → Increase in MPT → Increase in TFP growth → Increase in GDP growth → Increase in learning opportunity → Further learning efforts → Sustainable co-evolution,
(f) Confronting a post-information society (ubiquitous society), swell of Japan's institutional MOT adapting to new requirements in a ubiquitous society leads high-technology firms to transformation of technopreneurship resulting in the bi-polarization trajectory,
(g) OIR (operating income to R&D) substitution for R&D intensity by means of effective integration of innovation initiative and external acquisition efforts through cumulative market learning would be the key strategy corresponding to such a bi-polarization, and
(h) Institutional shift from co-existence or co-adaptation of homogeneous firms to co-evolution between heterogeneous firms would be essential.

References

1. IMD, The World Competitiveness Yearbook (IMD, Lausanne, annual issues)
2. C. Watanabe and R. Kondo, Institutional elasticity towards IT waves for Japan's survival, Technovation 23, No. 4 (2003) 307–320
3. The Economist, Japan Flying Again, The Economist, Feb. 12 (2004)
4. The Economist, The Sun Also Rises, The Economist, Oct. 8 (2005)
5. C. Watanabe and K. Fukuda, National innovation ecosystem: the similarlity and disparity of Japan–US technology policy systems toward a service oriented economy, Journal of Services Research 6, No. 1 (2006) 159–186
6. C. Chen and C. Watanabe, Diffusion, substitution and competition dynamism inside the ICT market: a case of Japan, Technological Forecasting and Social Change 73, No. 6 (2006) 731–759
7. K. Moriyama, The Transformation of Technopreneurial Structure in Japan and US High-technology Firms Confronting a Paradigm Shift to a Post-information Society (Master Thesis, Tokyo Institute of Technology, Tokyo, 2006) (*in Japanese*)
8. OECD, OECD Information Technology Outlook 2000 (OECD, Paris, 2000)
9. Council on Competitiveness, Innovative America (Council on Competitiveness, Washington, DC, 2004)
10. P. Kennedy, The Rise and Fall of the Great Powers (Unwin Hyman, London, 1988)
11. OECD, The New Economy: Beyond the Hype (OECD, Paris, 2001)
12. US DOC, Digital Economy 2000 (DOC, Washington, DC, 2000)
13. C. Watanabe, R. Kondo, N. Ouchi and H. Wei, A substitution orbit model of competitive innovations, Technological Forecasting and Social Change 71, No. 4 (2004) 365–390
14. C. Watanabe, R. Kondo, N. Ouchi, H. Wei and C. Griffy-Brown, Institutional elasticity as a significant driver of IT functionality development, Technological Forecasting and Social Change 71, No. 7 (2004) 723–750
15. A. Ohmura and C. Watanabe, Cross-products technology spillover in inducing a self-propagating dynamism for the shift to a service oriented economy: lessons from high-performance fine ceramics, Journal of Services Research 6, No. 2 (2006) 145–179
16. C. Watanabe and B. Asgari, Impacts of functionality development on the dynamism between learning and diffusion of technology, Technovation 24, No. 8 (2004) 651–664
17. Ministry of Internal Affairs and Communication of Japan (MIC), White Paper on Information and Communication in Japan (MIC, Tokyo, 2005)

Chapter 9
Technological Diversification: Strategic Trajectory Leading to an Effective Utilization of Potential Resources in Innovation: *A Case of Canon*

Abstract Under the paradigm shift from an industrial society to an information society in the 1990s, contrary to the decrease in operating income to sales in Japan's electrical machinery firms, only Canon demonstrated its increasing trend. This contrasting performance corresponds to Canon's another contrast with respect to increasing technological diversification while reverse trends in other electrical machinery firms. These contrasts suggest us that Canon's technological diversification strategy can be the source of high level of operating income to sales.

Prompted by this hypothetical view, this chapter attempts to elucidate a mechanism of Canon's technological diversification with special attention to its contribution to high level of operating income to sales by means of an effective utilization of potential resources in innovation. Comparative empirical analyses are conducted focusing on the consequence of technological diversification and development trajectory in Japan's leading electrical machinery firms over the last two decades.

Reprinted from *Technological Forecasting and Social Change* 72, No. 1, C. Watanabe, J. Y. Hur and K. Matsumoto, Technological Diversification and Firm's Techno-Economic Structure: An Assessment of Cannon's Sustainable Growth Trajectory, pages: 11–27, copyright (2005), with permission from Elsevier.

9.1 Introduction

Under the paradigm shift from an industrial society to an information society in the 1990s, contrary to the decrease in operating income to sales in Japan's electrical machinery firms, only Canon demonstrated its increasing trend. This contrasting performance corresponds to its consistent efforts to develop technological diversification over the last four decades.

While almost all electrical machinery firms decreased such diversification efforts in the 1990s, Canon's contrasting structure provides a constructive suggestion for survival strategy to electrical machinery firms suffering from income stagnation amidst megacompetition in an information society.

C. Watanabe, *Managing Innovation in Japan: The Role Institutions Play in Helping or Hindering how Companies Develop Technology*, DOI: 10.1007/978-3-540-89272-4_9,

Prompted by this observation, this paper attempts to elucidate Canon's dynamism leading to a virtuous cycle trajectory between technological diversification and high level of income structure by means of a comparative empirical analysis on the consequence of technological diversification and development trajectory in Japan's leading electrical machinery firms over the last two decades.

The significance of technological diversification has shifted to the relevance with assimilation of spillover technology while central issue of business diversification has moved to a prosperous business field. This shift was triggered by the technology distance concept postulated by Griliches [1]. Following this concept, Jaffe [2] developed new concepts of technological proximity and technological position. An empirical analysis was conducted by Goto and Suzuki [3, 4] utilizing R&D diversification function as a proxy of assimilation of spillover technology in Japan's electrical machinery industry.

Focus of research then shifted to firm's capability to utilize spillover technology for technological diversification. The concept of adaptive capacity was introduced by Cohen and Levintal [5, 6] to explain the capability adapting spillover technology to indigenous technology. Prompted by Cohen and Levintal, Watanabe et al. [7] developed the concept of assimilation capacity consisting of the ability for the perception, selection and incorporation of spillover technology.

While the importance of technological diversification has increased amidst mega-competition spurred by informatization and economic globalization in the 1990s, Japanese firms concentrated on their core business due to the economic stagnation after the bursting of the bubble economy [8]. Servas [9], Rajan et al. [10] and Scharfstein and Stein [11] postulated that research focus of the business diversification should shift to the evaluation of cost expansion accompanied with diversification from the viewpoint of excessive value approach. Therefore, the significant role of technological diversification is how to overcome this problem by constructing a virtuous cycle for an effective utilization of potential resources in innovation.

To date, while a number of studies have analyzed the significance of firm's technological diversification strategy, none has analyzed the dynamism of diversification, development trajectory, marginal productivity of technology, and subsequent income structure.

Thus, the focuses of this paper are three folds: first, measurement of technological diversification trajectory in leading electrical machinery firms based on the relationship between technology spillover and R&D diversification, second, structural analysis of R&D, technology, production structure as sources of technological diversification and income structure, and third, elucidation of the mechanism of technological diversification leading to high income structure.

Section 9.2 reviews theoretical background of R&D diversification and assimilation of spillover technology. Section 9.3 measures the extent of technological diversification. Section 9.4 analyzes the contribution of technological diversification to high income structure. Section 9.5 elucidates factors governing a virtuous cycle between inducing factors of R&D investment and technological diversification. Section 9.6 briefly summarizes new findings, their policy implications and future works.

9.2 R&D Diversification and Assimilation of Spillover Technology

9.2.1 Increasing Dependency on Spillover Technology

Amidst megacompetition, stimulated by the advancement of IT and consequent globalizing economy, the dependency on spillover technology has been significant [12,13]. In Japan, economic stagnation after the bursting of the bubble economy and consequent R&D stagnation have accelerated this trend [14].

Table 9.1 summarizes the trend in assimilated spillover technology from other firms in 24 Japan's leading electrical machinery firms over the period 1980–1998. As clearly observed in Fig. 9.1, not a few firms demonstrate their significance in the dependency on spillover technology in the 1990s. These trends demonstrate the significance of effective utilization of spillover technology in a global context.

a *DAST*: Dependency of Assimilated Spillover Technology

$$= \frac{Z \cdot T_{\mathrm{s}}}{T_{\mathrm{i}} + Z \cdot T_{\mathrm{s}}},$$

where T_{i}, indigenous technology stock; T_{s}, spillover technology pool, $T_{\mathrm{s}} = \sum_{j=1}^{24} T_j \quad (j \neq i)$; and assimilation capacity $Z = \frac{1}{1+\frac{\Delta T_{\mathrm{s}}/T_{\mathrm{s}}}{\Delta T_{\mathrm{i}}/T_{\mathrm{i}}}} \cdot \frac{T_{\mathrm{i}}}{T_{\mathrm{s}}}$ [7].

b Melco, Mitsubishi Electric Corporation; MEW, Matsushita Electric Works, Ltd.; JVC, Victor Company of Japan, Ltd.; and JRC, Japan Radio Co., Ltd.

9.2.2 Technology Distance, Technological Proximity and Technological Position

9.2.2.1 Technology Distance

Griliches [1] postulated that knowledge stock levels of firms depend on not only their indigenous R&D investment but also knowledge stock developed by other firms. He postulated that the latter effects the spillover of technology as depicted by the following equation:

$$Y_i = BX_i^{1-\gamma} K_i^{\gamma} K_{ai}^{\mu},$$

where Y_i, production of i industry; B, scale factor; X_i, labor and capital; K_i, indigenous knowledge stock; and K_{ai}, spillover knowledge stock.

$$K_{ai} = \sum_j w_{ij} K_j,$$

where w_{ij}, weight function enabling the spillover from j to i industry.

Table 9.1 Trends in dependency on assimilated spillover technology – 24 Japan's leading electrical machinery firms (1980–1998)

	Canon	Matsu-shita	NEC	Hitachi	Toshiba	Fujitsu	Melco	Sony	Sharp	Sanyo	MEW	JVC	Fuji	Kyo-cera	Oki	Pioneer	Alps	Casio	Rohm	Aiwa	Yoko-gawa	JRC	Meiden-sha	Kokusai
1980	0.29	0.29	0.34	0.35	0.36	0.33	0.35	0.35	0.33	0.27	0.35	0.32	0.36	0.44	0.33	0.35	0.34	0.34	0.37	0.31	0.31	0.35	0.31	0.20
1981	0.30	0.30	0.34	0.35	0.35	0.33	0.34	0.35	0.33	0.27	0.34	0.32	0.36	0.42	0.33	0.34	0.34	0.33	0.37	0.31	0.31	0.34	0.31	0.22
1982	0.31	0.30	0.34	0.35	0.35	0.33	0.34	0.34	0.33	0.27	0.33	0.33	0.35	0.40	0.33	0.34	0.34	0.33	0.36	0.31	0.32	0.34	0.31	0.24
1983	0.32	0.31	0.34	0.34	0.34	0.33	0.34	0.34	0.33	0.28	0.33	0.33	0.34	0.39	0.33	0.33	0.34	0.33	0.36	0.31	0.32	0.34	0.31	0.26
1984	0.32	0.31	0.34	0.34	0.34	0.33	0.34	0.34	0.33	0.29	0.33	0.34	0.34	0.38	0.33	0.33	0.34	0.33	0.36	0.31	0.32	0.34	0.31	0.28
1985	0.32	0.32	0.34	0.34	0.34	0.33	0.34	0.34	0.33	0.30	0.33	0.34	0.34	0.37	0.33	0.33	0.34	0.33	0.35	0.32	0.32	0.34	0.31	0.29
1986	0.33	0.32	0.34	0.34	0.34	0.34	0.33	0.34	0.33	0.30	0.32	0.33	0.34	0.37	0.33	0.33	0.34	0.33	0.35	0.32	0.33	0.34	0.31	0.30
1987	0.32	0.31	0.34	0.34	0.34	0.33	0.34	0.34	0.33	0.28	0.33	0.33	0.34	0.38	0.33	0.33	0.34	0.33	0.36	0.31	0.32	0.34	0.31	0.26
1988	0.33	0.32	0.34	0.34	0.34	0.34	0.33	0.33	0.34	0.28	0.32	0.33	0.33	0.36	0.33	0.32	0.33	0.35	0.35	0.31	0.33	0.34	0.32	0.32
1989	0.34	0.34	0.34	0.33	0.33	0.34	0.33	0.34	0.33	0.34	0.32	0.34	0.32	0.29	0.33	0.32	0.33	0.31	0.32	0.33	0.34	0.33	0.33	0.35
1990	0.34	0.34	0.33	0.32	0.33	0.35	0.33	0.35	0.31	0.39	0.32	0.32	0.32	0.31	0.32	0.33	0.25	0.34	0.30	0.35	0.35	0.32	0.34	0.38
1991	0.35	0.34	0.30	0.32	0.32	0.37	0.33	0.35	0.31	0.38	0.32	0.32	0.32	0.35	0.34	0.32	0.26	0.30	0.34	0.39	0.35	0.31	0.30	0.36
1992	0.35	0.33	0.29	0.32	0.32	0.37	0.34	0.35	0.32	0.36	0.32	0.32	0.31	0.37	0.36	0.33	0.25	0.27	0.36	0.41	0.34	0.30	0.32	0.36
1993	0.35	0.33	0.28	0.33	0.32	0.36	0.34	0.37	0.33	0.36	0.33	0.31	0.31	0.38	0.36	0.34	0.12	0.30	0.38	0.39	0.33	0.31	0.33	0.36
1994	0.35	0.33	0.28	0.33	0.32	0.36	0.32	0.38	0.34	0.36	0.34	0.30	0.26	0.39	0.33	0.35	0.19	0.32	0.36	0.38	0.33	0.32	0.33	0.37
1995	0.37	0.33	0.26	0.32	0.33	0.36	0.30	0.38	0.35	0.38	0.35	0.28	0.31	0.35	0.32	0.37	0.26	0.34	0.36	0.41	0.32	0.33	0.33	0.37
1996	0.39	0.32	0.25	0.32	0.34	0.33	0.30	0.38	0.37	0.39	0.37	0.27	0.32	0.28	0.25	0.39	0.23	0.36	0.38	0.43	0.31	0.34	0.34	0.37
1997	0.41	0.31	0.27	0.31	0.34	0.31	0.30	0.38	0.38	0.39	0.37	0.27	0.30	0.25	0.22	0.38	0.14	0.36	0.40	0.44	0.28	0.36	0.34	0.33
1998	0.40	0.31	0.30	0.29	0.34	0.32	0.30	0.39	0.37	0.37	0.36	0.29	0.29	0.32	0.26	0.36	−0.37	0.33	0.41	0.44	0.29	0.35	0.31	0.35

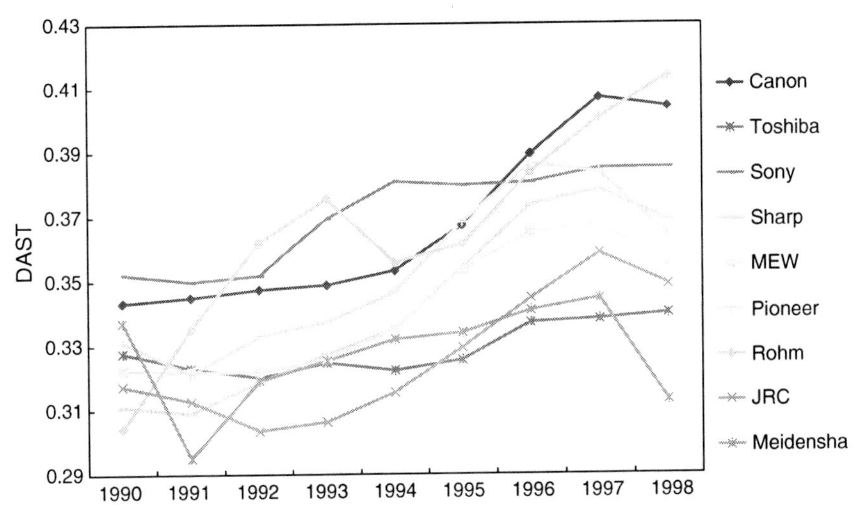

Fig. 9.1 Trend in dependency on assimilated spillover technology of Japan's leading electrical machinery firms (1980–1998)

Griliches demonstrated that effective utilization of spillover technology is subject to this weight function that is solely dependent on the technology distance between donor and host firms. This has two important implications.

(a) First, those firms very close to host firm share similar products as well as R&D, thereby developed such infrastructure as manufacturing facilities, R&D resources and marketing networks leading to easy utilization and absorption of spillover technology.
(b) Second, strong possibility to come across timely necessitated technology is expected.

Thus, he demonstrated the significance of effective utilization of spillover technology for the improvement of firm's collective knowledge level and its dependency on its efforts to shortening technology distance.

9.2.2.2 Technological Proximity and Technological Position

Jaffe [2], stimulated by Griliches's suggestion, postulated a new concept of technological position based on the R&D similarity of host and donor firms measured by technological proximity as practical technological distance between the two firms.

Given the technological position of firm i, the technological distance between firms i and j, P_{ij} can be depicted by the following equation:

$$P_{ij} = \frac{F_i \cdot F_j'}{[\,(F_i \cdot F_i')(F_j \cdot F_j')\,]^{1/2}}, \tag{9.1}$$

where F_i, distribution of R&D expenditure in firm i $F_i = (F_{i1} \cdots F_{ij} \cdots F_{in})$; F_i', transpose vector; and F_{ij}, ratio of R$D expenditure in the field of j expensed by firm i.

This distance varies between 0 and 1 and it approaches 1 as the proximity increases.

Utilizing this equation, assimilated spillover technology of firm i out of total R&D investment in donor firm j, $[R_i]_j$ can be measured by the following equation:

$$[R_i]_j = R_j \cdot P_{ij}. \tag{9.2}$$

9.2.3 R&D Diversification and Technological Distance

On the basis of the concept of technology distance, assimilation capacity which plays a key role in effective utilization of spillover technology, is governed by the proximity of technological position of both donor and host sides.

Technological distance could be maximized when the proximity of technological position corresponds each other. This proximity depends on the similarity of R&D activities between the donor and the host.

This is because when the host conducts similar R&D activities of the donor it can realize the trend, existence and absorbability of spillover technology for the effective and advantageous introduction, absorption and development of spillover technology.

Diversification of R&D is a strategy in line with this direction and actively employed during the course of the 1980s in Japan (MITI) [15].

Given R_{ij} is the R&D expenditure in the field of j out of R&D expenditure of sector i, Ri, the ratio of R&D investment initiated by sector i, D_i can be depicted by the following equation:

$$D_i = R_{ii}/R_i. \tag{9.3}$$

Thus, diversification ratio D_{ni} can be expressed as follows:

$$D_{ni} = 1 - D_i = 1 - R_{ii}/R_i. \tag{9.4}$$

This ratio can be also measured as a function of Herfindahl Index (HHI) as follows:

$$D_{ni} = f(\text{HHI}), \ \text{HHI} = 1 - \sum P_i^2, \tag*{(9.4)'}$$

where $P_i = R_{ii}/\Sigma R_{ii}$.

Utilizing the diversification ratio, the technological distance in (9.1) can be depicted as follows:

$$P_{ij} = \frac{\sum_k \frac{R_{ik}}{R_i} \cdot \frac{R_{jk}}{R_j}}{\left[\sum_j \frac{R_{ij}{}^2}{R_i} \cdot \sum_i \frac{R_{ji}{}^2}{R_j} \right]^{1/2}}. \tag{9.5}$$

Thus, technology assimilated by sector i out of R_j initiated by sector j can be depicted as the follows by means of the product of R_j and P_{ij} [3,4]:

$$[R_i]_j = R_j \cdot P_{ij} = F(D_{ni}). \tag{9.6}$$

Since D_{ni} is a function of HHI as depicted in (9.4)', assimilated spillover technology $([R_i]_j)$ in (9.6) can be depicted as a function of HHI as follows:

$$[R_i]_j = H(\text{HHI}). \tag{9.6'}$$

Equation (9.6)' enumerates a correlation between technological diversification represented by HHI and assimilated spillover technology.

9.3 Measurement of Technological Diversification

9.3.1 *Measurement of Canon's Technology Development Trajectory*

Canon's technology development trajectory can be represented by the trend in patents registered in the USA [16]. This trend demonstrates very strong correlation with that of its sales as illustrated in Fig. 9.2.

Based on these observations, Canon's technology development trajectory can be measured utilizing patents extraction method classifying 20,252 Canon's patents registered in the USA over the period 1976–2000 into 33 technological fields.

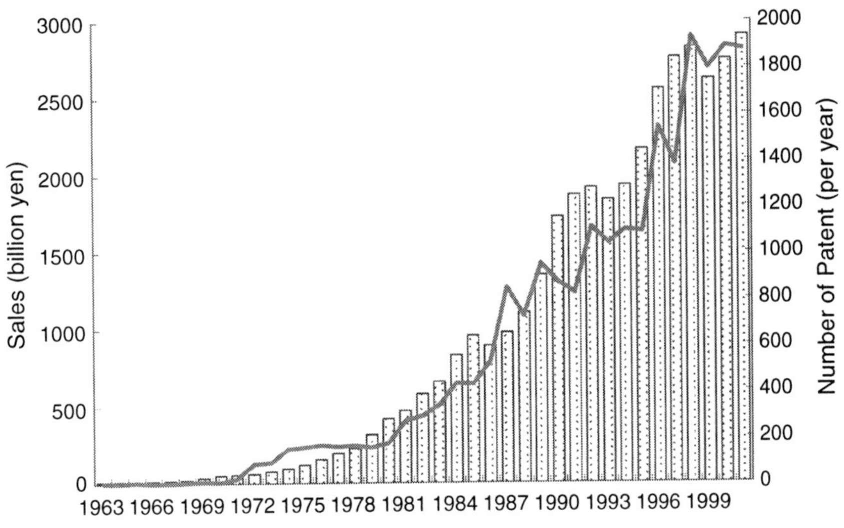

Fig. 9.2 Trends in number of patents registered in the USA and sales of Canon (1963–2001)
Source: Canon (2002)

Figure 9.3 illustrates these trends by integrating into ten technological fields. Tabulated number of patents demonstrated in Fig. 9.3 as a proxy of Canon's technology development trajectory by technological fields is summarized in Table 9.2.

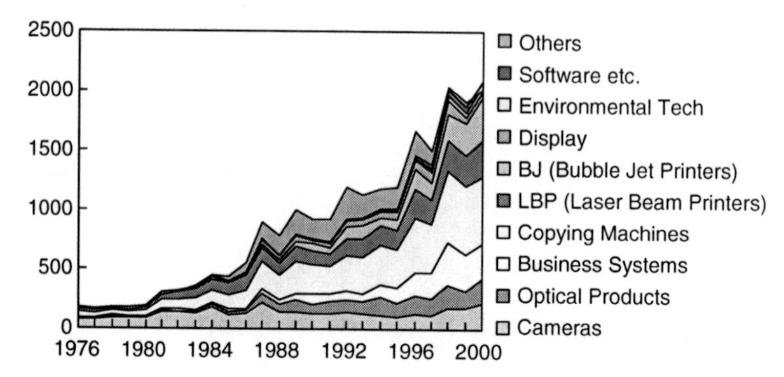

Fig. 9.3 Canon's technology development trajectory by means of patents registered in the USA (1976–2000)

Table 9.2 Canon's patents numbers registered in the USA (1976–2000)

	Cameras	Optical products	Business systems	Copying machines	LBP	BJ	Display	Environmental tech	Software	Others
1976	74	7	10	47	35	1	0	1	0	0
1977	75	5	16	35	21	6	3	0	0	3
1978	95	4	11	40	23	1	0	0	0	0
1979	88	5	15	34	27	2	4	1	0	0
1980	85	5	12	49	33	3	1	0	0	0
1981	135	21	12	81	55	3	4	1	0	0
1982	136	11	14	80	66	5	0	3	0	0
1983	126	15	14	96	85	6	1	1	4	6
1984	178	23	14	106	86	23	13	3	0	0
1985	116	26	19	125	93	7	10	4	3	39
1986	127	23	12	155	134	11	9	3	3	80
1987	216	74	38	235	118	27	32	5	8	141
1988	143	68	35	199	100	32	33	5	4	169
1989	146	94	54	270	132	38	45	9	13	201
1990	127	79	83	253	107	64	30	8	7	169
1991	134	100	66	232	98	49	36	8	6	190
1992	143	101	76	292	146	103	46	7	8	276
1993	127	102	66	310	155	108	58	7	6	186
1994	115	154	105	325	175	62	53	13	13	169
1995	103	116	126	318	174	84	68	12	18	171
1996	128	159	192	455	247	160	79	17	26	203
1997	121	138	219	413	205	138	87	31	36	111
1998	184	182	365	600	259	216	115	39	56	28
1999	176	141	311	576	253	269	62	35	45	37
2000	224	199	300	565	302	357	59	49	39	0

Table 9.3 Patents extraction formula for the analysis of Canon's technological development

Field	Patents extraction formula
Whole fields	AN/Canon and ISD/$/$/1998
Cameras	AN/Canon and ((ICL/G02B0$ or ICL/G03B0$ or ICL/G03C0$ or ICL/G03D0$) andnot (ICL/G03B027/$ or ICL/G03B042/$)) and ISD/$/$/1998
Optical products	AN/Canon and (ICL/G03B042/$ or ICL/G03C005/16 or ICL/H05G0$ or ICL/H01J035/$ or ICL/A61B0$ or ICL/G03F0$ or ICL/H01L0$ or ABST/ray or ABST/semiconduct$) and ISD/$/$/2000
Business systems	AN/Canon and (ICL/G06F0$ or ICL/G11C0$ or ICL/G06K0$ or ABST/calculat$ or ABST/computer$) and ISD/$/$/2000
Copying machines	AN/Canon and (ABST/copy$ or ABST/photocopy$ or ABST/toner$ or ABST/photoconductor$ or ABST/develop$ ICL/G03B027/$ or ICL/H04N$ or ABST/fax$ or ABST/scan$) and ISD/$/$1978
LBP	AN/Canon and ICL/G03G0$ and ISD/$/$/2000
BJ	AN/Canon and ICL/B41J0$ and ISD/$/$/2000
Display	AN/Canon and (ICL/G02F001/13$ or ICL/C09K019/$ or (ABST/liquid$ and ABST/crystal$) or ABST/LCD) and ISD/$/$/2000
Environmental technology	AN/Canon and (ICL/C12$ or ABST/ecology$ or ABST/recycl$ or ABST/biolog$ or ABST/biochem$ or ABST/environment$ or ABST/wast$ or ABST/remed$) and ISD/$/$/2000
Software	AN/Canon and (ABST/softwar$ or ABST/program$ or ABST/manag$) and ISD/$/$/2000

Ten technological fields are (1) cameras, (2) optical products, (3) business systems (information and communication equipment), (4) copying machines, (5) LBP, (6) BJ, (7) liquid crystal display, (8) environment technology, (9) software technology, (10) miscellaneous and those patents extraction method is summarized in Table 9.3.

9.3.2 Measurement of Technological Diversification

Utilizing Canon's technological development trajectory classified into 33 technological fields, and depending on (9.4)', trends in Canon's technological diversification represented by Herfindahl index over the period 1978–1998 are measured. Figure 9.4 illustrates these trends that demonstrate chronology of Canon's technological diversification strategy as summarized in Table 9.4.

Utilizing both trends in Canon's Herfindahl index (Fig. 9.4) and assimilated spillover technology (Table 9.1) and based on (9.6)', correlation analysis between technological diversification and dependency on assimilated spillover technology over the period 1980–1998 is conducted.

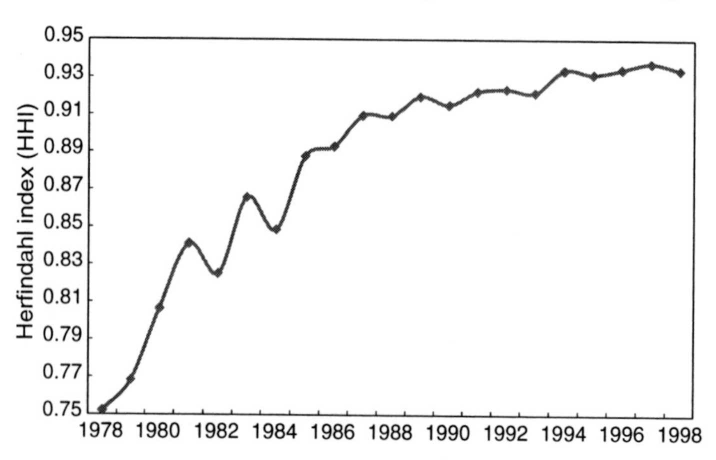

Fig. 9.4 Trends in Canon's technological diversification (1978–1998)

Table 9.4 Chronology of Canon's technological diversification strategy

Period	Management decision	Organization	Products	Technology
1960	*Embryo period of diversification* 45 Declaration for revival 47 Canon Camera Co. Ltd. *Diversification planning period*	55 New York branch 57 Distribution post in Europe	40 X-ray camera 59 Synchroreader	Optics, precision machinery +Electronics
1970	62 First five year plan *Declaration of diversification* 67 Slogan "cameras in right hand, business products in left hand" 69 Canon Inc.	62 Canon Latin America 62 Product research dept. 64 Project base system 69 Central R&D institute 70 Production base in Taiwan	61 Canonnet 61 Microfilm system 64 Calculator 68 Original electro-photography 70 Copying machine	+Electrophotography recording
1980	*Diversification development period* 76 Premier company plan	72 Production base in Germany 74 Production base in California 77 Business division system	75 LBP 76 AE-1, facsimiles 79 LBP-10 80 Japanese word processor	+Laser recording
1990	85 Tie-up with HP in computer business 88 Restart with global company plan	81 Corporate development center 83 Production base in France 84 Production of copying machines in China 88 R&D center in UK 90 Overseas labs (America, Australia, France)	81 BJ 82 PC-10/20 84 LBP-CX 85 Optical card R/W 90 BJ-10	87 Optics, precision machinery, production, electronics, recording, memory, new recording, software, communication, system, biotechnology

(Continued)

Table 9.4 (continued)

Period	Management decision	Organization	Products	Technology
2000	98 Management renovation committee	92 Canon Chemicals 93 Ecology R&D center 96 Fuji-Susono research park 98 Joint venture with Beijing university 99 Ayase office (semiconductor)	91 FLCD 92 BJC-full color bubble jet printer 95 Optical card system 96 Advanced photo system camera 99 Surface-conduction electron emitter display 00 Compact digital video camcorder	21 key technologies Precision recording, optics, instrument & control, production process, image processing, memory, opto-electronics, semiconductor, display, etc.

Source: Canon (2001) [25]

$$\ln \text{DAST} = -0.990 + 0.976 \ln \text{HHI} - 0.065 D_{87} \quad \text{adj.}R^2 = 0.922 \quad \text{DW} = 1.61, \tag{9.7}$$
$$(-90.04) \; (13.78) \qquad (-3.28)$$

where DAST, dependency on assimilated spillover technology; HHI, Herfindahl index; and D_{87}, dummy variable (starting year of the bubble economy).

Equation (9.7) indicates extremely high statistical significance and demonstrates significant correlation between DAST and HHI in Canon over the period 1980–1998.

A postulate on the relationship between R&D diversification and technology distance as reviewed in Sect. 9.2.3 suggests that (9.7) represents the correlation between technological diversification and assimilated spillover technology not only in Canon but also common to firms consisting technology spillover pool (in case of the analysis in Table 9.1, 24 electrical machinery firms). Therefore, based on this correlation and utilizing assimilated spillover technology measured in Table 9.1, the trend in technological diversification in each respective firm in electrical machinery industry as listed in Table 9.1 can be measured.

Results of the analysis on trends in technological diversification of 15 leading electrical machinery firms out of 24 firms measured by this approach are demonstrated in Fig. 9.5.

Looking at Fig. 9.5, we note that while technological diversification has activated in the 1980s, almost all firms decreased their diversification in the 1990s as a consequent of the shift of business strategy to selection and concentration. These trends correspond to the preceeding empirical analysis [8].

Noteworthy is that in this trend only Canon has maintained further diversification. Figure 9.6, that analyzes the correlation between technological diversification and R&D intensity demonstrates clear positive correlation between them. Canon that leads the highest technological diversification ratio together with Sony demonstrates the highest R&D intensity.

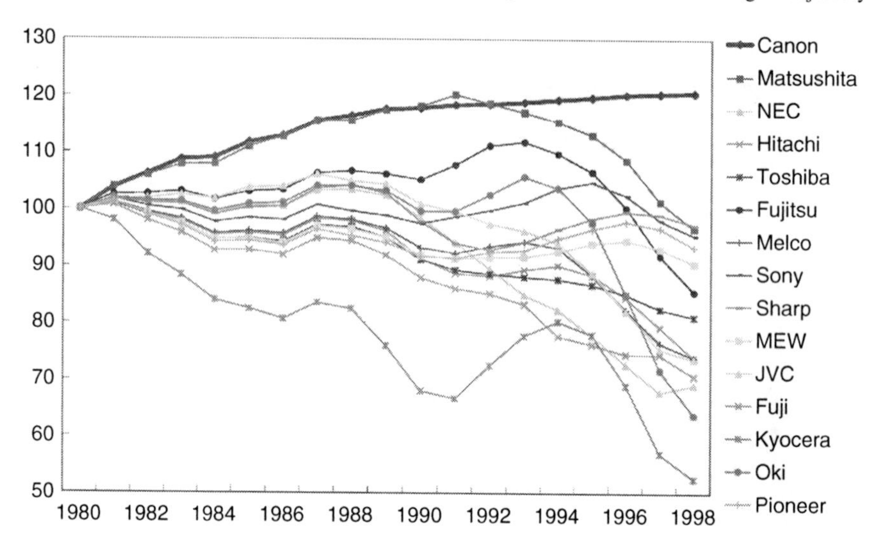

Fig. 9.5 Trends in technological diversification of 15 leading electrical machinery firms (1980–1998) – Herfindahl index HHI 1980 = 100, 3 years moving average

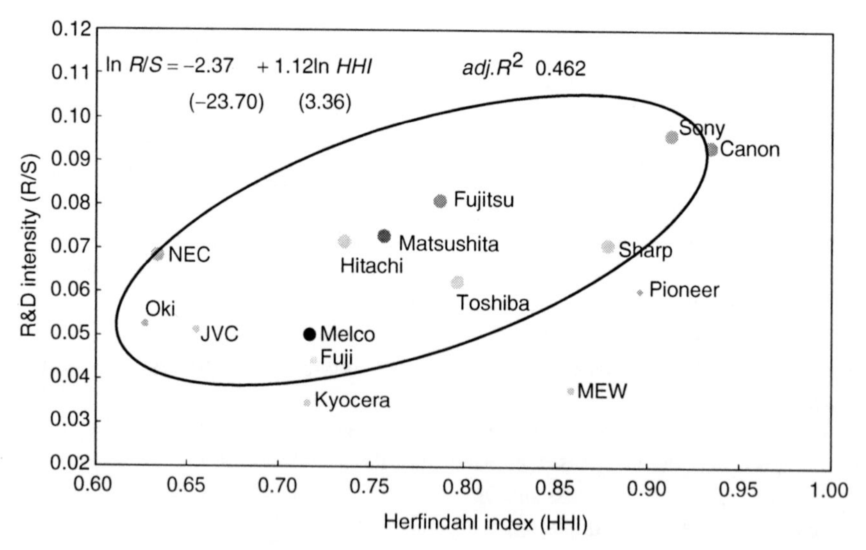

Fig. 9.6 Correlation between technological diversification and R&D intensity in 15 leading electrical machinery firms (1995–1998)

[a] *R/S* R&D intensity 1990 fixed price, R&D expenditure per sales

9.4 Contribution of Technological Diversification to High Income Structure

9.4.1 Contribution to Operating Income to Sales (OIS)

9.4.1.1 Trends in OIS in Ten Leading Electrical Machinery Firms

Figure 9.7 as well as Table 9.5 demonstrates the trend in OIS of ten leading electrical machinery firms (top ten firms with respect to sales volume excluding Sanyo[1]) over the period 1982–1998.

Looking at these figure and table, we note that while Canon declined its OIS in the middle of 1980s and in the beginning of the 1990s, it recovered immediately and maintained high level of OIS.

It is noteworthy that while other electrical machinery firms decreased their OIS in the 1990s, only Canon increased its OIS and has maintained high level of OIS.

Table 9.6 summarizes trends in Canon's sales composition on which its OIS depends. As illustrated in Fig. 9.8, its selling cost has demonstrated dramatic decrease, particularly in the 1990s, which is considered the main source of maintaining high level of OIS compared with those of other firms [17].

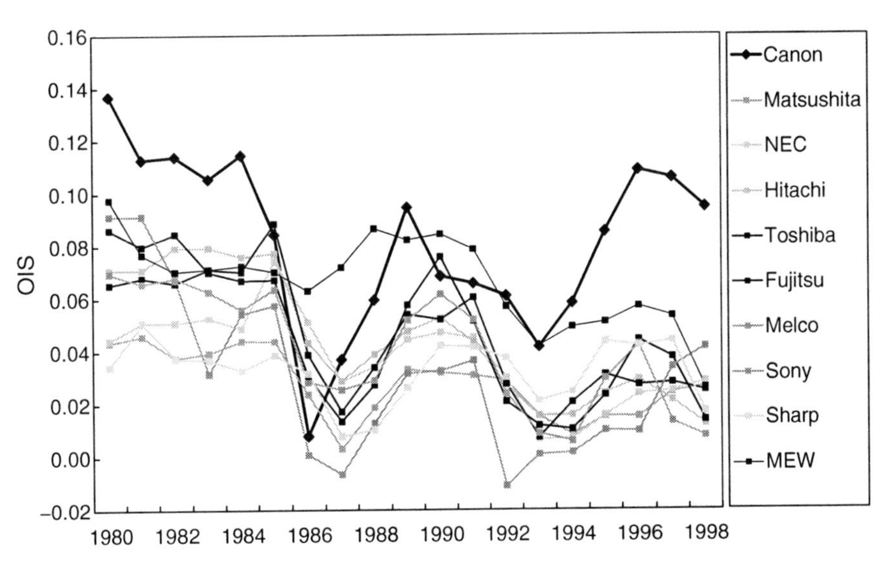

Fig. 9.7 Trends in OIS of ten leading electrical machinery firms (1980–1998)

[1] Sanyo (10th sales volume in 1998) was excluded as it was not necessarily R&D intensive firm in the early 1980s.

Table 9.5 Comparison of OIS of ten leading electrical machinery firms (1980–1998)

	1980–1986	1987–1990	1991–1994	1995–1998
Canon	0.10	0.06	0.06	0.10
Matsushita	0.04	0.02	0.02	0.02
NEC	0.05	0.04	0.02	0.02
Hitachi	0.07	0.04	0.03	0.02
Toshiba	0.07	0.04	0.02	0.03
Fujitsu	0.07	0.04	0.03	0.03
Melco	0.06	0.04	0.02	0.02
Sony	0.06	0.02	0.01	0.02
Sharp	0.04	0.02	0.03	0.04
MEW	0.07	0.08	0.06	0.05

1980–1986, after the second energy crisis and before the bubble economy; 1987–1990, during the bubble economy; 1991–1994, after the bursting of the bubble economy and before the economic stagnation; and 1995–1998, during the severe economic stagnation

Table 9.6 Trends in Canon's sales composition (1980–1998): %

	1980	1981	1982	1983	1984	1985	1986	1987	1988	1989	1990
Cost of goods manufactured	51.0	54.0	52.7	52.6	54.4	54.8	59.8	60.6	58.0	57.0	55.8
Selling cost	9.0	9.2	9.1	9.2	9.0	9.1	9.3	8.7	8.5	8.6	8.7
General administrative expenditure	25.7	26.6	26.7	27.0	26.5	26.8	27.5	25.9	25.6	25.8	26.2
Operating income	14.3	10.3	11.5	11.2	10.1	9.3	3.4	4.7	7.9	8.6	9.2
Sales	100	100	100	100	100	100	100	100	100	100	100

	1991	1992	1993	1994	1995	1996	1997	1998
Cost of goods manufactured	56.2	58.6	61.4	61.2	61.3	59.3	57.3	57.4
Selling cost	9.1	8.4	8.3	8.2	7.8	7.9	8.1	8.2
General administrative expenditure	27.3	25.3	25.1	24.9	23.7	24.2	24.7	25.1
Operating income	7.4	7.7	5.3	5.7	7.2	8.6	10.0	9.3
Sales	100	100	100	100	100	100	100	100

9.4.1.2 Analysis of Contributing Factors to OIS

Contributing factors to OIS of electrical machinery firms are analyzed. It is generally postulated that OIS in high-technology firms are governed by economic situation, exchange rate as well as firms' techno-preneurship strategy [14]. Therefore, the following function depicting governing factors of OIS is formulated by incorporating diversification strategy as techno-preneurship strategy.

$$\mathrm{OIS} = A \cdot (\mathrm{CI})^{b}(\mathrm{YR})^{c}(\mathrm{HHI})^{d}\, e^{\lambda t}, \tag{9.8}$$

where A, scale factor; CI, economic situation (composite index); YR, exchange rate (\/\$); t, time trend; and b, c, d and λ, elasticity of each factor.

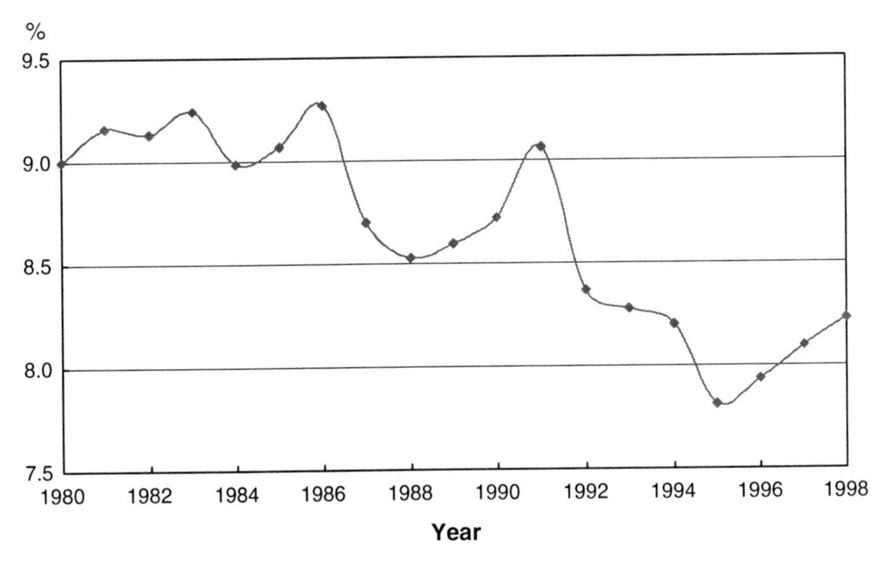

Fig. 9.8 Trends in the ratio of selling cost out of Canon's total sales (1980–1998)

Utilizing (9.8), governing factors of OIS of ten leading electrical machinery firms are analyzed by (9.9).

$$\ln \text{OIS} = a + b \ln \text{CI} + c \ln \text{YR} + d_1 D_{80-86} \ln(\text{HHI}_i) + d_2 D_{87-90} \ln(\text{HHI}_i) + d_3 D_{91-94} \ln(\text{HHI}_i) + d_4 D_{95-98} \ln(\text{HHI}_i) + \lambda t, \quad (9.9)$$

where D is a coefficient dummy variable corresponding to the classification of the period depending on Japan's economic situation as indicated in Table 9.5.

$$(D_{80-86}\text{: } 1980\text{–}1986 = 1 \text{ others} = 0 \; D_{87-90}\text{: } 1987\text{–}1990 = 1 \text{ others} = 0)$$
$$D_{91-94}\text{: } 1991\text{–}1994 = 1 \text{ others} = 0 \; D_{95-98}\text{: } 1995\text{–}1998 = 1 \text{ others} = 0$$

Results of the analysis are summarized in Table 9.7 which demonstrates conspicuously high Canon's technological diversification elasticity to OIS, suggesting that Canon's technological diversification contributed significantly to OIS and effects of economic situation (CI) as well as exchange rate (YR) were not so significant.

OIS is governed by the share of cost of goods manufactured (COGM), selling cost as well as general administrative expenditure. Since sales of R&D intensive electrical machinery firms depend largely on innovative products, the selling cost greatly influences their OIS.

However, given the new innovative products are attractive enough, a firm can increase its sales without depending on increase in selling cost.

As analyzed in Fig. 9.8, Canon decreased dramatically its dependency on selling cost in the 1990s and this seems to be the major source of maintaining high level of OIS.

Table 9.7 Factors contributing to OIS in ten leading electrical machinery firms (1980–1998)

$$\ln \text{OIS} = a + b \ln \text{CI} + c \ln \text{YR} + d_1 D_{80-86} \ln(\text{HHI}_i) + d_2 D_{87-90} \ln(\text{HHI}_i) + d_3 D_{91-94} \ln(\text{HHI}_i) + d_4 D_{95-98} \ln(\text{HHI}_i) + \lambda t$$

	a	b	c	d_1	d_2	d_3	d_4	λ	adj.R^2	DW
Canon	−1.47	0.48	−0.54	*1.89*	*52.35*	*55.76*	*49.63*	−0.13	0.955	2.78
	(−0.35)	(0.73)	(−1.22)	*(0.53)*	*(3.44)*	*(3.48)*	*(3.17)*	(−1.83)		
Matsushita	−24.37	3.28	−1.52	*7.27*	*8.87*	*−1.80*	*−3.70*	−0.16	0.889	2.00
	(3.03)	(1.80)	(2.88)	*(1.36)*	*(1.65)*	*(0.57)*	*(−1.60)*	(−1.44)		
NEC	−21.94	2.11	−1.66	*1.04*	*8.42*	*5.06*	*2.90*	0.10	0.850	1.69
	(−3.94)	(1.78)	(−2.81)	*(0.18)*	*(1.67)*	*(1.99)*	*(1.69)*	(1.91)		
Hitachi	−17.59	1.96	−1.12	*1.53*	*4.47*	*3.62*	*2.73*	–	0.900	1.61
	(−5.17)	(2.84)	(−4.13)	*(0.54)*	*(1.70)*	*(2.59)*	*(3.32)*			
Toshiba	−18.86	2.25	−1.81	*−4.33*	*−12.29*	*15.59*	*7.00*	−0.13	0.891	2.45
	(3.75)	(2.40)	(−2.50)	*(−0.72)*	*(−1.68)*	*(1.72)*	*(0.86)*	(1.36)		
Fujitsu	−20.14	2.53	−1.12	*0.136*	*6.02*	*3.40*	*1.25*	–	0.933	2.56
	(−5.98)	(3.39)	(−2.96)	*(0.04)*	*(1.61)*	*(0.74)*	*(1.37)*			
Melco	−29.01	4.41	−1.17	*6.60*	*13.38*	*8.72*	*3.72*	–	0.630	1.59
	(3.25)	(2.35)	(−1.62)	*(0.95)*	*(1.73)*	*(1.76)*	*(1.82)*			
Sony	−72.37	7.61	−6.91	*48.82*	*30.07*	*22.08*	*−4.75*	–	0.788	2.32
	(−5.03)	(2.50)	(−6.35)	*(3.64)*	*(2.10)*	*(1.83)*	*(−0.57)*			
Sharp	−11.51	0.39	−0.85	*−17.37*	*−16.15*	*−12.85*	*−21.47*	–	0.567	2.19
	(−1.92)	(0.29)	(−1.17)	*(−2.53)*	*(−2.57)*	*(−3.77)*	*(−4.38)*			
MEW	−10.70	2.41	0.48	*4.14*	*10.17*	*−3.04*	*2.41*	–	0.884	2.62
	(−5.14)	(5.88)	(2.03)	*(2.86)*	*(5.41)*	*(−2.79)*	*(4.04)*			

Table 9.8 Comparison of decreasing effects of COGM, selling cost and general administrative expenditure by means of technological diversification (1980–1998)

$$\ln X = a + b \ln \text{HHI}$$

X	a	b	adj.R^2	DW
C/S	−0.46	0.99	0.642	1.22
	(−22.67)	(5.76)		
Sell/S	−2.54	−0.85	0.496	1.02
	(−109.81)	(−4.33)		
GA/S	−1.41	−0.47	0.232	0.94
	(−65.03)	(−2.54)		

S, sales; C, COGM; sell, selling cost; GA, general administrative expenditure; and HHI, Herfindahl index

Suggested by these observations, the effects of technological diversification to decrease in COGM, selling cost and general administrative expenditure are analyzed and the results are summarized in Table 9.8, which suggests that both selling cost and general administrative expenditure, particularly selling cost, can be decreased as technological diversification is developed.

Based on these findings, the following scheme can be envisaged: increase in technological diversification → increase in functionality of new innovative goods → decrease in dependency on selling cost → increase in OIS.

9.4.2 Technological Diversification and Marginal Productivity of Technology

9.4.2.1 Analysis of Development Trajectory of Leading Electrical Machinery Firms

Diffusion and development trajectory of high-technology products are actually quite similar to the contagion process of an epidemic disease [18] and exhibits S-shaped growth. This process is well modeled by the following simple logistic growth function (epidemic model) which was first introduced by Vehrulst in 1845 [19]:

$$\frac{\mathrm{d}\Sigma Y(t)}{\mathrm{d}t} = a\sum Y(t)\cdot\left(1-\frac{\Sigma Y(t)}{K}\right), \tag{9.10}$$

where $\Sigma Y(t)$, cumulative production of high-technology goods at time t; K, carrying capacity; and a, coefficients governing the diffusion velocity as follows:

$$a = \frac{\Delta\Sigma Y(t)}{\Sigma Y(t)}\Big/\left(1-\frac{\Sigma Y(t)}{K}\right).$$

By developing (9.10), the following epidemic model depicting technological trajectory can be obtained:

$$\sum Y(t) = \frac{K}{1+\mathrm{Exp}(-at-b)}, \tag{9.11}$$

where b, coefficient.

Since high-technology products can be considered as the crystal of technology stock and the sales of high-technology firms are proportional to the development of their high-technology products [14] and technology stock increase as time goes by, the epidemic model of (9.11) can be expressed by (9.12).

$$\sum S(T) = \frac{\overline{\Sigma S}}{1+\mathrm{Exp}(-aT-b)}, \tag{9.12}$$

where $\overline{\Sigma S}$, carrying capacity of cumulative production; and T, technology stock.

Given the rate of obsolescence of high-technology products ρ, increasing rate at initial stage g, cumulative sales can be approximated as $\sum S \approx S/(\rho+g)$. Therefore, under the condition that $\rho+g$ is stable, (9.12) can be approximated by the following equation:

$$S(T) = \frac{\bar{S}}{1+\mathrm{Exp}(-aT-b)}. \tag{9.13}$$

The development trajectory expressed by (9.13) assumes that the level of carrying capacity is constant through the development process of technology. However, in reality, the interaction between technology and institutions induces a structural change and enhance the level of carrying capacity in line with increase of technology stock and development of new functionality [20]. Therefore, $\bar{S}(T)$ is also an epidemic model enumerated as follows:

$$\bar{S}(T) = \frac{\bar{S}_K}{1+\mathrm{Exp}(-a_K T - b_K)}, \tag{9.14}$$

where a_K and b_K, coefficients; and $\bar{S}_K$, ultimate carrying capacity.

The solution of the differential (9.10) under the condition (9.14) can be obtained as follows:

$$S(T) = \frac{\bar{S}_K}{1+\mathrm{Exp}(-aT-b)+\frac{a}{a-a_K}\mathrm{Exp}(-a_K T - b_K)}. \tag{9.15}$$

Utilizing (9.15), development trajectories of Japan's leading electrical machinery firms over the period 1980–1998 are analyzed. Results of analysis are summarized in Table 9.9.

Table 9.9 compares estimations both by a simple logistic growth function (9.13) and a logistic growth function within a dynamic carrying capacity (9.15). Comparing AIC (Akaike Information Criteria) we note that Table 9.9 suggests that the estimations by a logistic growth function within a dynamic carrying capacity demonstrates statistically more significant except Matsushita which stands the top level of sales and most matured stage in Japan's electrical machinery firms.

As Table 9.9 demonstrates extremely high statistical significance, (9.15) can be considered to represent development trajectory of high-technology firms driven by their technology stock.

9.4.2.2 Measurement of Marginal Productivity of Technology

Based on the foregoing analysis, the marginal productivity of technology of high-technology firms can be measured by (9.15). Taking partial differentiation of (9.15) with respect to technology stock T, the following equation depicting marginal productivity of technology can be measured.

$$\begin{aligned}\frac{\partial S}{\partial T} &= \frac{\bar{S}_K\left(a\,\mathrm{Exp}(-aT-b)+\frac{aa_K}{a-a_K}\mathrm{Exp}(-a_K T-b_K)\right)}{\left(1+\mathrm{Exp}(-aT-b)+\frac{a}{a-a_K}\mathrm{Exp}(-a_K T-b_K)\right)^2}\\ &= a\,S(T)\left(1-\frac{S(T)}{\bar{S}(T)}\right) = a\,S(T)\left(1-\frac{1}{\mathrm{FD}}\right),\end{aligned} \tag{9.16}$$

where $\mathrm{FD} \equiv \bar{S}(T)/S(T)$: degree of functionality development.

In the process of diffusion of high-technology products, the ratio of carrying capacity to the level of diffusion represents the extent of functionality development [21,22] and $\bar{S}(T)/S(T)$ in (9.16) represents this ratio (which is defined as the degree of functionality development).

Therefore, (9.16) suggests that the marginal productivity increases as the functionality development increases. In addition, the marginal productivity of technology is proportional to diffusion velocity (a) as well as sales.

Table 9.9 Estimation results for the development trajectories of Japan's leading Electrical machinery firms (1980–1998)

	K	a	b	a_K	b_K	adj.R^2	DW	AIC
Canon	2,529	0.607×10^{-2}	−3.04			0.995	0.94	4,124
	(14.50)	(13.60)	(−31.00)					
	21,929	0.900×10^{-2}	−5.54	0.116×10^{-2}	−3.22	0.999	1.18	2,773
	(1.18)	(4.97)	(−6.09)	(2.94)	(−3.41)			
Matsushita	6972	0.158×10^{-2}	−2.12			0.993	0.89	56,456
	(11.71)	(6.53)	(−12.43)					
	6,972	0.935×10^{-1}	−4.00	0.158×10^{-2}	−2.11	0.983	0.89	161,490
	(11.73)	(8.59)	(−4.00)	(6.54)	(−12.46)			
NEC	6431	0.169×10^{-2}	−3.03			0.977	0.46	171,220
	(4.48)	(4.27)	(−12.32)					
	8,719	0.853×10^{-2}	−9.06	0.110×10^{-2}	−2.38	0.968	0.67	54,621
	(5.75)	(3.15)	(−3.50)	(5.51)	(−8.36)			
Hitachi	7,527	0.138×10^{-2}	−2.11			0.985	0.52	89,084
	(2.97)	(3.22)	(−13.45)					
	13,808	0.465×10^{-2}	−3.89	0.801×10^{-3}	−2.00	0.914	0.62	38,528
	(3.07)	(2.27)	(−2.65)	(4.23)	(−4.83)			
Toshiba	5,409	0.244×10^{-2}	−2.55			0.987	0.59	54,601
	(11.73)	(7.03)	(−13.69)					
	19,338	0.848×10^{-2}	−7.16	0.828×10^{-3}	−2.50	0.935	1.11	44,854
	(2.01)	(3.09)	(−4.05)	(5.75)	(−4.92)			
Fujitsu	5,599	0.172×10^{-2}	−2.27			0.965	0.42	137,420
	(2.47)	(3.13)	(−8.66)					
	28,453	0.981×10^{-2}	−6.49	0.821×10^{-3}	−3.22	0.936	0.64	30,662
	(2.60)	(3.73)	(−5.54)	(8.57)	(−7.70)			
Melco	3,781	0.346×10^{-2}	−1.93			0.983	0.60	33,740
	(21.25)	(8.83)	(−14.60)					
	16,898	0.740×10^{-2}	−4.30	0.699×10^{-3}	−2.13	0.916	0.93	15,581
	(2.34)	(5.78)	(−7.65)	(5.79)	(−4.46)			
Sony	3972	0.301×10^{-2}	−2.46			0.992	0.58	15,768
	(3.18)	(3.58)	(−14.44)					
	17,764	0.979×10^{-2}	−5.33	0.139×10^{-2}	−3.06	0.991	0.65	15,646
	(2.27)	(2.75)	(−4.94)	(4.93)	(−5.79)			
Sharp	1,932	0.600×10^{-2}	−1.80			0.939	0.45	20,740
	(8.52)	(4.70)	(−10.99)					
	3,869	0.126×10^{-1}	−2.96	0.201×10^{-2}	−1.32	0.958	0.57	20,724
	(1.78)	(2.70)	(−2.81)	(2.13)	(−1.64)			

Upper two lines indicate estimation by a simple logistic growth function (9.13) and the lower two lines indicate that of by a logistic growth function within a dynamic carrying capacity (9.15)

The marginal productivity of technology in leading electrical machinery firms measured by (9.16) is summarized in Table 9.10.

Looking at Table 9.10, we note that Canon's marginal productivity of technology was the lowest among leading electrical machinery firms compared during the period of 1980–1986, it increased and ranked top level from the late 1980s.

Figures 9.9 and 9.10 analyze the correlation between technological diversification and marginal productivity of technology as well as diffusion velocity in leading electrical machinery firms. Both figures demonstrate clear positive correlation suggesting that technological diversification contributed to the increase of marginal productivity of technology and velocity of technology diffusion.

Table 9.10 Trends in marginal productivity of technology in leading electrical machinery firms (1980–1998)

	1980–1986	1987–1990	1991–1994	1995–1998
Canon	1.72	3.02	3.67	4.77
Matsushita	2.25	2.72	2.25	1.45
NEC	3.64	2.41	2.45	2.62
Hitachi	2.74	2.42	2.48	2.62
Toshiba	4.68	2.81	2.50	2.88
Fujitsu	3.17	1.90	1.99	2.65
Melco	3.93	2.98	2.01	1.97
Sony	2.79	2.56	2.66	3.28
Sharp	3.09	2.80	2.06	1.85

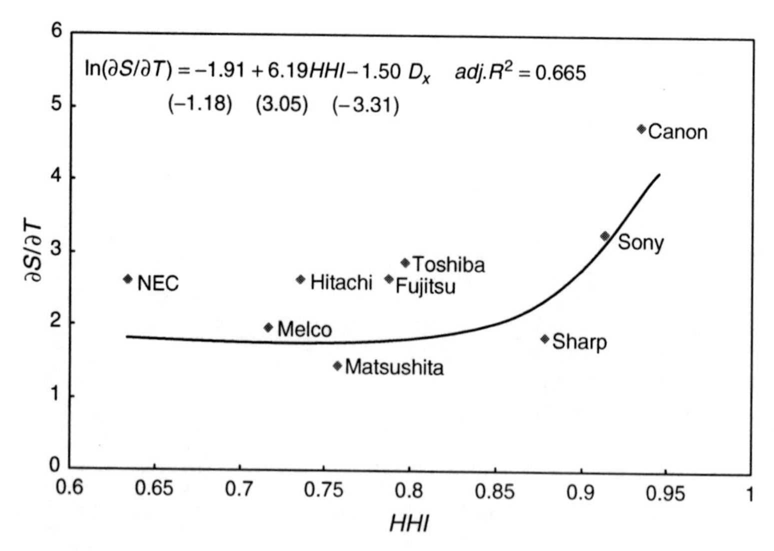

Fig. 9.9 Correlation between technological diversification and marginal productivity of technology (1995–1998 average)

[a] D_x, Dummy variable for Matsushita and Sharp

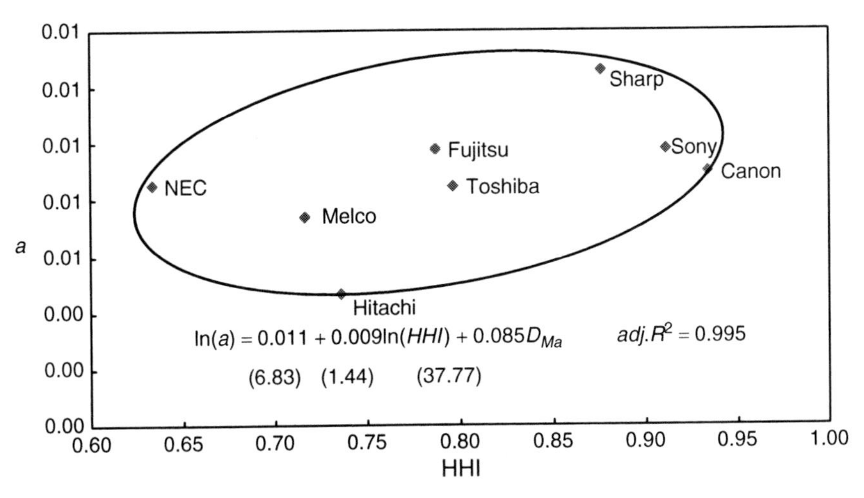

Fig. 9.10 Correlation between technological diversification and velocity of technology diffusion (1995–1998 average)

[a] $a = \frac{\Delta S(T)}{S(T)}\left(1 - \frac{S(T)}{\bar{S}(T)}\right)^{-1}$, velocity of technology diffusion

[b] D_{Ma}, Dummy variable for Matsushita

9.4.2.3 Measurement of Functionality Development

As analyzed before, Canon's marginal productivity of technology has increased and maintained top level in the 1990s. Equation (9.16) suggests that this can be attributed to the functionality development together with high velocity of technology diffusion.

Analyses in Table 9.6, Fig. 9.8 and Table 9.8 on the contribution of selling cost decrease to Canon's OIS also suggest a significant contribution of the functionality development by means of technological diversification efforts to its high level of OIS.

Since the degree of functionality development can be depicted by the following equation, trends in functionality development in leading electrical machinery firms can be measured utilizing the estimated results summarized in Table 9.9:

$$\mathrm{FD} = \frac{\bar{S}(T)}{S(T)} = \frac{\frac{\bar{S}_K}{1+\mathrm{Exp}(-a_K T - b_K)}}{\frac{\bar{S}_K}{1+\mathrm{Exp}(-aT-b)+\frac{a}{a-a_K}\mathrm{Exp}(-a_K T - b_K)}}$$

$$= \frac{1+\mathrm{Exp}(-aT-b)+\frac{a}{a-a_K}\mathrm{Exp}(-a_K T - b_K)}{1+\mathrm{Exp}(-a_K T - b_K)}. \quad (9.17)$$

Table 9.11 summarizes the result of the functionality development measured by an epidemic function developed by (9.15) in leading electrical machinery firms over the period 1980–1998, which demonstrates Canon has maintained high level of the functionality development.

Table 9.11 Trends in functionality development of leading electrical machinery firms (1991–1998)

	1991–1994	1995–1998
Canon	1.53	1.29
Matsushita	1.01	1.00
NEC	1.14	1.04
Hitachi	1.20	1.09
Toshiba	1.13	1.07
Fujitsu	1.16	1.07
Mitsubishi	1.13	1.07
Sony	1.18	1.14
Sharp	1.18	1.09

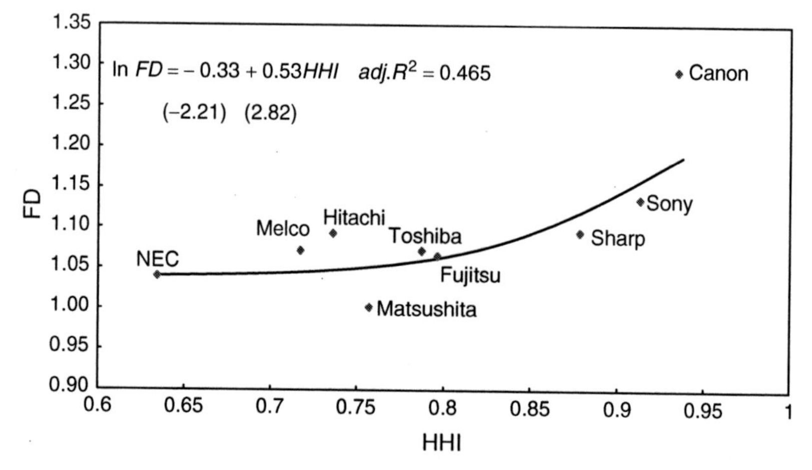

Fig. 9.11 Correlation between technological diversification and functionality development in leading electrical machinery firms (1995–1998 average)

Figure 9.11 analyzes the correlation between technological diversification and functionality development measured by an epidemic model depicted in (9.15) which illustrates clear positive correlation demonstrating that technological diversification contributed to functionality development.

Table 9.12 summarizes the chronology of Canon's functionality development over the last three decades. Looking at the table, we note that Canon has developed variety of technologies based on its indigenous technology and created new functionality similar to chain reaction in its innovation process leading to construction of the following virtuous cycle: technological diversification → development of new functionality → creation of variety of technologies → further development of new functionality.

Synchronizing results of these correlations as well as (9.16), the relationship between technological diversification, marginal productivity of technology, velocity of technology diffusion and functionality development can be systemized in Fig. 9.12.

Table 9.12 Chronology of Canon's new functionality development over the last three decades

Year	New functionality development
1970	Japan's first plain-paper copying machine "NP-1100" by original electro-photography technology
	Japan's first mask aligner "PPC-1" by precision optical technology
1976	"AE-1" camera with a built-in microcomputer
	World's first nonmydriatic retinal camera "CR-45NM"
1978	World's first screen processing copying machine "NP-8500"
1979	"LBP-10" by applying semiconductor laser
1981	World's first bubble jet printing technology development
1982	World's first personal copying machine "PC-10/20"
1984	Digital laser copying machine system "NP-9030"
	World's smallest laser beam printer "LBP-8/CX"
1985	World's first bubble jet printer "BJ-80"
1986	Still video system
1988	World's highest resolution CCD built-in still video camera "RC-760"
1990	Note book size bubble jet printer "BJ-10"
1991	World's first FLCD (Ferro-electric liquid crystal display)
1996	APS (Advanced photo system) double zoom camera "IXY"
1999	Joint development of "SED (Surface-conduction electron-emitter displays)"
2000	Compact and card sized digital camera "IXY DIGITAL"

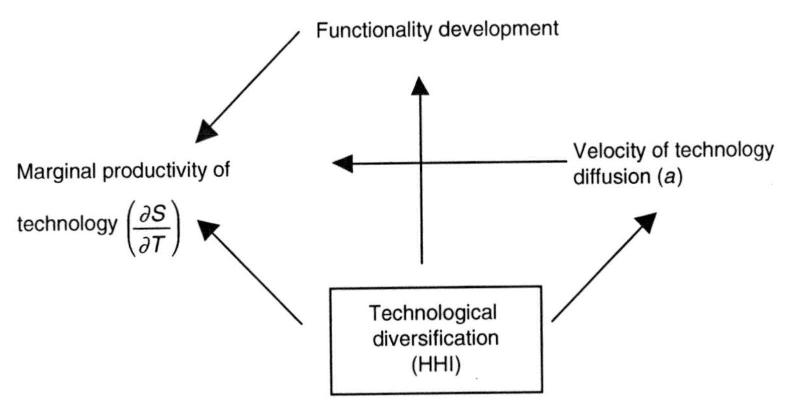

Fig. 9.12 Relationship between technological diversification, marginal productivity of technology, velocity of technology diffusion and functionality development

9.4.3 Technological Diversification and Increase in Total Factor Productivity

As reviewed in Sect. 9.4.2, sales in high-technology firms are proportional to the volume of high-technology products as crystal of technology stock. Thus, growth rate of total factor productivity (TFP) can be depicted by the following equations:

$$S = F(X, T), \tag{9.18}$$

where S, sales; X, labor (L), capital (K); intermediate input (M) and T, technology stock.

Table 9.13 Comparison of increasing rate of total factor productivity in leading electrical machinery firms (1980–1998)

	1980–1986	1987–1990	1991–1994	1995–1998
Canon	0.284	0.417	0.396	0.443
Matsushita	0.200	0.250	0.178	0.106
NEC	0.620	0.287	0.224	0.179
Hitachi	0.281	0.257	0.234	0.188
Toshiba	0.406	0.235	0.196	0.179
Fujitsu	0.517	0.254	0.232	0.213
Melco	0.290	0.220	0.127	0.099
Sony	0.353	0.331	0.306	0.314
Sharp	0.278	0.225	0.165	0.131

Let $\frac{dS}{dt} \equiv \Delta S$, $\frac{dX}{dt} \equiv \Delta X$ and $\frac{dT}{dt} \equiv \Delta T \approx R$ R: R&D investment

$$\frac{\Delta S}{S} = \sum_{X=L,K,M} \left(\frac{\partial S}{\partial X} \cdot \frac{X}{S}\right) \frac{\Delta X}{X} + \left(\frac{\partial S}{\partial T} \cdot \frac{T}{S}\right) \frac{\Delta T}{T} \tag{9.19}$$

$$\frac{\Delta \text{TFP}}{\text{TFP}} = \frac{\Delta S}{S} - \sum_{X=L,K,M} \left(\frac{\partial V}{\partial X} \cdot \frac{X}{S}\right) \frac{\Delta X}{X} = \left(\frac{\partial S}{\partial T} \cdot \frac{T}{S}\right) \frac{\Delta T}{T} \approx \frac{\partial S}{\partial T} \cdot \frac{R}{S}$$

$$\therefore \frac{\Delta \text{TFP}}{\text{TFP}} = \frac{\partial S}{\partial T} \cdot \frac{R}{S}. \tag{9.20}$$

Synchronizing marginal productivity of technology measured by (9.16) and R&D intensity (Fig. 9.6), trend in growth rate of TFP in leading electrical machinery firms over the period 1980–1998 can be measured by (9.20). The results are summarized in Table 9.13.

Looking at Table 9.13, we note that Canon has maintained the highest level of TFP growth rate from the late 1980s corresponding to increase in its marginal productivity of technology as demonstrated in Table 9.10. Equation (9.20) suggests that this high level of TFP growth rate can be attributed to high level of marginal productivity of technology and R&D intensity.

Integrating these results, Fig. 9.12 which illustrates the relationship between technological diversification, marginal productivity of technology, velocity of technology diffusion and functionality development can be developed to a sophisticated dynamism incorporating impacts of R&D intensity, TFP growth and sales increase as illustrated in Fig. 9.13.

9.4.4 Technological Diversification and Internal Rate of Return to R&D Investment

Given the lead time from R&D investment to commercialization m, rate of obsolescence of technology stock ρ and current discount rate r, the equilibrium between

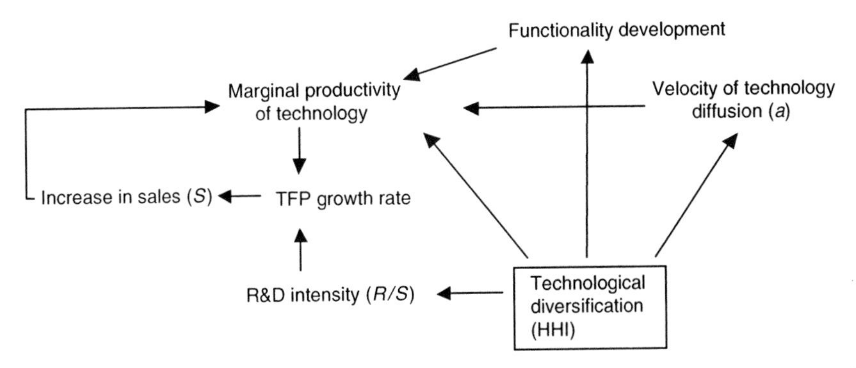

Fig. 9.13 Dynamism between technological diversification, functionality development, tfp growth rate and sales increase

Table 9.14 Comparison of internal rate of return to R&D investment in leading electrical machinery firms (1980–1998)

	1980–1986	1987–1990	1991–1994	1995–1998
Canon	0.518	0.774	0.887	1.061
Matsushita	0.630	0.725	0.645	0.485
NEC	0.858	0.670	0.684	0.728
Hitachi	0.719	0.672	0.691	0.729
Toshiba	1.000	0.738	0.694	0.775
Fujitsu	0.787	0.570	0.594	0.733
Melco	0.901	0.766	0.599	0.601
Sony	0.726	0.697	0.723	0.842
Sharp	0.776	0.737	0.609	0.577

one unit of R&D investment and present value of consequent benefit can be depicted by the following (9.21) [23]:

$$e^{mr} = \int_0^{\infty} \frac{\partial S}{\partial T} e^{-(\rho+r)t} dt = \frac{\partial S}{\partial T} \Big/ (\rho + r). \tag{9.21}$$

By developing Taylor series of the left-hand side to the first order, (9.22) can be obtained.

$$1 + mr = \frac{\partial S}{\partial T} \Big/ (\rho + r). \tag{9.22}$$

Solving (9.22), internal rate of return (IRR) to R&D investment r can be depicted by the following equation:

$$r = \frac{-(1+m\rho) + \sqrt{(1+m\rho)^2 - 4m(\rho - \frac{\partial S}{\partial T})}}{2m}. \quad \because r \geqslant 0 \tag{9.23}$$

Substituting $\partial S/\partial T$ by (9.16), the internal rate of return to R&D investment in leading electrical machinery firms can be measure. Results of the measurement are summarized in Table 9.14.

Looking at this table, we note that trend in Canon's IRR to R&D investment is similar to the trend in its marginal productivity of technology which is summarized in Table 9.10. While Canon's IRR to R&D investment was the lowest among leading electrical machinery firms during the period of 1980–1986, it increased and ranked top level from the late 1980s.

9.5 Factors Inducing a Virtuous Cycle Between R&D Investment and Technological Diversification

9.5.1 Inducing Factors of Technological Diversification

The foregoing analyses suggest that Canon's high level of OIS in the 1990s particularly when majority of electrical machinery firms decreased in their OIS can be attributed to the sustainable technological diversification effort. This effort was conspicuous as majority of electrical machinery firms shifted to converge by selection and concentration. Only Canon has maintained technological diversification.

Stimulated by these noteworthy observations, Table 9.15 analyzes inducing factors of technological diversification and compares elasticity of R&D intensity, marginal productivity of technology, TFP growth rate and IRR to R&D investment to technological diversification focusing on Canon as it demonstrates high level of technological diversification.

Looking at Table 9.15, we note that IRR to R&D investment demonstrates the highest statistical significance and provides noteworthy inducement of technological diversification.

Generally, in order to develop diversification, certain resources are indispensable for firms [24]. Therefore, focusing on nine leading electrical machinery firms, correlation analyses are conducted by incorporating sales as the resource for diversification together with IRR to R&D investment to identify inducing factors to technological diversification.

Table 9.15 Comparative analysis of inducing factors of Canon's technological diversification (1980–1998)

$\ln \text{HHI} = a + b \ln X_{-1}$ X_{-1} $\frac{R}{S}$, $\frac{\partial S}{\partial T}$, $\frac{\Delta \text{TFP}}{\text{TFP}}$, IRR at time $t-1$

X	a	b	adj.R^2	DW
$\frac{R}{S}$	0.04 (1.62)	−1.06 (−5.86)	0.663	0.71
$\frac{\partial S}{\partial T}$	−0.19 (−17.43)	0.03 (8.74)	0.816	1.50
$\frac{\Delta \text{TFP}}{\text{TFP}}$	−0.24 (−11.40)	0.38 (6.73)	0.722	1.88
IRR	−0.23 (−17.19)	0.17 (9.80)	0.848	1.88

Table 9.16 Inducing factors in leading electrical machinery firms (1980–1998)

$$\ln \text{HHI} = a + b \ln \text{IRR}_{-1} + c \ln S_{-1}$$

	a	b	c	adj.R^2	DW
Canon	−0.203	0.057	0.019	0.905	1.87
	(−2.31)	(1.91)	(1.56)		
Matsushita	0.270	0.459	−0.028	0.850	1.50
	(2.11)	(9.65)	(−1.78)		
NEC	1.733	−0.500	−0.273	0.661	0.51
	(4.72)	(−1.61)	(−4.83)		
Hitachi	1.120	−0.639	−0.186	0.818	1.01
	(5.13)	(−1.75)	(−6.44)		
Toshiba	1.166	−0.134	−0.170	0.903	2.04
	(8.41)	(−2.19)	(−7.74)		
Fujitsu	0.391	−0.424	−0.091	0.489	0.66
	(2.30)	(−3.81)	(−3.58)		
Melco	0.709	0.245	−0.107	0.677	1.00
	(1.32)	(1.27)	(−1.37)		
Sony	−0.208	−0.240	0.008	0.487	1.93
	(−2.05)	(−2.40)	(0.68)		
Sharp	0.170	−0.106	−0.050	0.355	1.35
	(1.01)	(−0.78)	(−1.60)		

The results of analyses are summarized in Table 9.16.

Although firms except Sony demonstrates statistical significance of sales as a resource inducing technological diversification, correlation between them demonstrates negative except Canon, which suggests the necessity of diversification increase as sales decrease. Only Canon increased technological diversification together with increase of sales and this is very noteworthy observation.

It is generally pointed out that firms with sufficient business resources in core business field possess high ability to diversify. However, sufficient business resources, ironically, suggest no need to diversify. Conversely, necessity to diversify is high when core business is stagnated. However, in that circumstance, the resources for diversification are limited. In this regard, there exists generally a paradox between the necessity of diversification and resources for diversification.

The sales is definitely a significant source for technological diversification. Table 9.16 demonstrates that seven firms demonstrating negative correlation between sales and technological diversification depend on technological diversification when their core businesses are stagnated. However, only Canon has challenged its technological diversification making full utilization of appropriate diversification resources when its core business increases and therefore technological diversification is not so urgent. This is considered the sources constructing a virtuous cycle between technological diversification and increase in OIS compared to other electrical machinery firms in the 1990s.

On the basis of the foregoing analyses, it has been identified that sources enabling Canon to maintain technological diversification even in the 1990s, can be attributed to the highest level of R&D investment supported by its higher level of IRR to

R&D investment as well as its continuing techno-preneurship strategy to appropriate resources for technological diversification while its core business increases.

9.5.2 Factors Governing Internal Rate of Return to R&D Investment

Stimulated by the inducing factors of technological diversification, focusing on Canon's IRR to R&D investment as primary contributor to technological diversification, governing factors of IRR to R&D investment are analyzed.

It is generally pointed out that IRR to R&D investment is governed by efforts to increase firm's R&D investment represented by R&D intensity and external factors such as exchange rate [23].

Therefore, an analysis is conducted taking R&D intensity (R/S), Yen exchange rate (YR), economic situation (CI), technological diversification (HHI) and time trend (t) by using the following equation:

$$\mathrm{IRR} = A\ (R/S)^b\ \ (\mathrm{YR})^c\ \ (\mathrm{CI})^d\ (\mathrm{HHI})^{De} e^{\lambda t}.$$

Table 9.17 summarizes the result of analysis which demonstrates that Canon's elasticity of technological diversification to IRR to R&D investment is higher than that of R&D intensity. The elasticity of economic situation (statistically non-significant) as well as exchange rate is statistically insignificant and their impacts on IRR to R&D investment are low.

9.5.3 Dynamism Leading to a Virtuous Cycle for Technological Diversification

On the basis of the foregoing analyses, we can conclude that the following cyclical dynamism as illustrated in Fig. 9.14, based on a virtuous cycle between technological diversification and IRR to R&D investment, which contributed to increase in OIS in the 1990s enabled Canon to increase and sustain high level of its OIS:

Table 9.17 Analysis on governing factors of IRR in Canon (1980–1998)

$\ln \mathrm{IRR} = a + b \ln R/S + c \ln \mathrm{YR} + d \ln \mathrm{CI} + D_1 e_1 \ln \mathrm{HHI} + D_2 e_2 \ln \mathrm{HHI} + \lambda t$

	a	b	c	d	e_1	e_2	λ	adj.R^2	DW
Coefficient	−123.03	0.55	−0.22	0.10	4.01	3.76	0.06	0.968	1.88
	(−3.96)	(1.89)	(−1.89)	(0.51)	(3.42)	(2.36)	(3.94)		

IRR, internal rate of return; R/S, R&D intensity R&D per sales; YR, Yen rate $/\; CI, composite index; D_1, D_2, dummy variable; D_1, 1980–1990 = 1, others 0, D_2, 1991–1998 = 1, others 0; HHI, Herfindahl index; and t, time trend

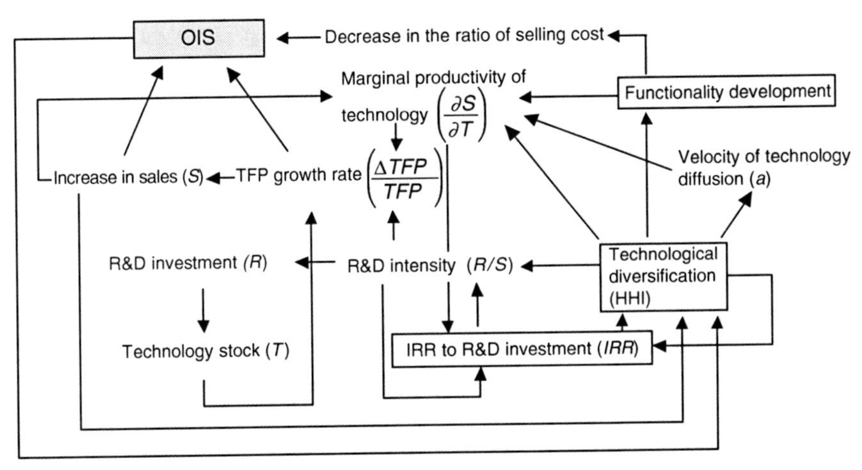

Fig. 9.14 A virtuous cycle enabling Canon to increase and sustain high level of OIS

(a) Supported by its consistent technological diversification strategy, Canon has succeeded to achieve successive new functionality development.
(b) This diversification contributes to improve marginal productivity of technology as well as IRR to R&D investment.
(c) Increased IRR to R&D investment induces R&D intensity, which, together with increased marginal productivity of technology, increases TFP growth leading to the effective increase in sales.
(d) These high productivities together with decrease in the ratio of selling cost attributed by functionality development contribute to increase in OIS.
(e) Increase in sales and OIS further encourages technological diversification strategy which further induces functionality development as well as high productivities leading to constructing a virtuous cycle between technological diversification and high level of OIS.

9.6 Conclusion

Prompted by conspicuous Canon's behavior with respect to technological diversification as well as high level of operating income to sales, this chapter attempted to elucidate its dynamism. Based on a theoretical relationship between assimilation capacity of spillover technology and diversification of R&D activities, new methodology to measure the extent of technological diversification of firms by means of assimilated spillover technology was developed.

Utilizing this methodology, technological diversification trend in Japan's leading electrical machinery firms was measured and its impacts on R&D intensity was analyzed. In addition, based on a logistic growth function within a dynamic carrying capacity incorporating technology stock, development trajectories of these firms were estimated, thereby marginal productivity of technology, functionality development, TFP growth rate, IRR to R&D investment and impacts of technological diversification on them were analyzed. All correlation analyses between them revealed that Canon demonstrated the highest significance from the late 1980s.

Inducing factors of technological diversification were analyzed and both IRR to R&D investment and sales were identified as sources of Canon's technological diversification.

Noteworthy is that while Canon's technological diversification increased as its sales increased, reverse trends were observed in other electrical machinery firms, which supports a postulate of diversification paradox. This demonstrates that Canon, against such paradox, sustained its technological diversification consistently by utilizing its sales increase fully.

These results demonstrate that Canon constructed a sophisticated virtuous cycle between high level of IRR to R&D investment, technological diversification and increase in operating income to sales. These observations provide noteworthy suggestions to firms, particularly electrical machinery firms amidst megacompetition as follows:

(a) Canon's technological diversification strategy is beyond the discussion on selection and concentration as well as diversification paradox leading to a new diversification theory in an information society.
(b) Canon's diversification strategy provides a constructive suggestion for effective utilization of spillover technology which is considered the key of competitiveness in an increasing trend in global technology spillover.
(c) Thus, Canon's technological diversification trajectory provides a constructive suggestion to best utilization of potential resources in innovation by constructing an effective virtuous cycle.

Future works should focus on the adaptability of Canon's business model based on such technological diversification strategy to other firms facing similar situation.

References

1. Z. Griliches, Issues in assessing the contribution of R&D to productivity growth, Bell Journal of Economics 10 (1979) 92–116
2. B. Jaffe, Technological opportunity and spillovers of R&D: evidence from firm's patents, profits, and market value, The American Economic Review 76, No. 5 (1986) 984–1001
3. A. Goto and K. Suzuki, R&D diversification and technology spillover effect, Economic Research 38, No. 4 (1987) 298–306 *(in Japanese)*
4. A. Goto and K. Suzuki, R&D capital, rate of return on R&D investment and spillover of R&D in Japanese manufacturing industries, The Review of Economics and Statistics 71, No. 4 (1989) 555–564
5. W.M. Cohen and D.A. Levinthal, Innovation and learning: the two faces of R&D, The Economic Journal 99 (1989) 569–596
6. W.M. Cohen and D.A. Levinthal, Absorptive capacity: a new perspective on learning and innovation, Administrative Science Quarterly 35, No. 1 (1990) 128–152

7. C. Watanabe, M. Takayama, A. Nagamatsu, T. Tagami and C. Griffy-Brown, Technology spillover as a complement for high-level R&D intensity in the pharmaceutical industry, Technovation 22, No. 4 (2002) 245–258
8. K. Gemba and F. Kodama, Diversification dynamics of Japanese industry, Research Policy 30, No. 8 (2001) 1165–1184
9. H. Servas, The value of diversification during the conglomerate merger wave, Journal of Finance 51 (1996) 1201–1225
10. R. Rajan, H. Servaes and Z. Luigi, The cost of diversity: the diversification and inefficient investment, Journal of Finance 55 (2000) 35–80
11. D. Scharfstein and J. Stein, The dark side of internal capital markets, divisional rent-seeking and inefficient investment, Journal of Finance 55 (2000) 2537–2564
12. OECD, Technology and Industrial Performance (OECD, Paris, 1997)
13. OECD, Technology, Productivity and Job Creation: Best Policy Practices (OECD, Paris, 1998)
14. C. Watanabe, B. Asgari and A. Nagamatsu, Virtuous cycle between R&D, functionality development and assimilation capacity for competitive strategy in Japan's high-technology industry, Technovation 23, No. 11 (2003) 879–900
15. Ministry of International Trade and Industry, "Trend and Task of Industrial Technology" (Research Institute of Economy, Trade and Industry, Tokyo, 1988) (*in Japanese*)
16. C. Watanabe, Y. Tsuji and C. Griffy-Brown, Patent statistics: deciphering a real versus a pseudo proxy of innovation, Technovation 21, No. 12 (2001) 783–790
17. K. Matsumoto, An Analysis of Technology Structure Inducing Canon's Diversification Strategy (Ph.D. Dissertation, Tokyo Institute of Technology, Tokyo, 2003) (*in Japanese*)
18. T. Modis, Prediction (Simon & Schnster, New York, 1992)
19. P.S. Meyer, Bi-logistic growth, Technological Forecasting and Social Change 47, No. 1 (1994) 89–102
20. P.S. Meyer and J.H. Ausbel, Carrying capacity: a model with logistically varying limits, Technological Forecasting and Social Change 61, No. 3 (1999) 209–214
21. F. Kodama, Innovation management in the emerging it environments, in J.A.D. Machuca and T. Mandakovic (eds.), POM Facing the New Millennium (Production and Operation Management Society, Sevilla, Spain, 2000)
22. C. Watanabe, R. Kondo, N. Ouchi, H. Wei and C. Griffy-Brown, Institutional elasticity as a significant driver of IT functionality development, Technological Forecasting and Social Change 71, No. 7 (2004) 723–750
23. C. Watanabe and K. Wakabayashi, The perspective of technometabolism and its insight into national strategies, Research Evaluation 6, No. 2 (1996) 69–76
24. T. Itami and T. Kagono, Introduction to Economics Seminar (Nihon Keizai Shimbun, Tokyo, 2001) (*in Japanese*)
25. Canon, THE CANON STORY 2001 (Canon, Tokyo, 2001)

Chapter 10
Japan's Coevolutionary Dynamism Between Innovation and Institutional Systems: *Hybrid Management Fusing East and West*

Abstract Contrary to its long-lasting economic stagnation during the "lost decade" in the 1990s, Japan is expected to "flying again." This anticipation can largely be attributed to the activation of Japan's indigenous virtuous cycle between technological innovation and economic development.

Despite many handicaps, Japan achieved a conspicuous technological advancement and subsequent productivity increase by devoting technology substitution for constrained production factors such as labor in the 1960s and energy in the 1970s. Such efforts enabled Japan to improve its institutional systems essential for its technological innovation, which in turn induced further innovation. Thus Japan constructed a sophisticated coevolutionary dynamism between innovation and institutional systems.

However, its economic stagnation in an information society in the 1990s demonstrates that this dynamism may stagnate if institutional systems cannot adapt to innovations.

Noteworthy surge in new innovation in recent years in leading edge activities of Japan's certain high-technology firms can be attributed to the coevolution between indigenous strength developed in an industrial society and effects of the cumulative learning from their competitors in an information society. This coevolution emerges as hybrid management by fusing "east" (indigenous strength) and "west" (learning from and corresponding to digital economy) leading to Japan's firms being more resilient against ubiquitous economy where seamless, on demand and open-sourcing are essential requirements.

Empirical analysis is focused on the elucidation of the coevolutionary domestication leading the noted hybrid management fusing east and west.

10.1 Introduction

Coevolutionary dynamism between innovation and institutional systems is decisive for an innovation driven economy.

C. Watanabe, *Managing Innovation in Japan: The Role Institutions Play in Helping or Hindering how Companies Develop Technology*, DOI: 10.1007/978-3-540-89272-4_10,

Innovation in institutional systems can be understood by means of a three-dimensional system consisting of:

(a) A national strategy and socioeconomic system
(b) An entrepreneurial organization and culture
(c) Historical perspectives

An innovation-driven economy may stagnate if institutional systems cannot adapt to innovations, and Japan's economy in the 1990s is one example.

Japan indigenously incorporates an explicit function which induces coevolutionary dynamism enabling it to achieve conspicuous performance in a virtuous cycle between innovation and rapid economic growth in the 1960s followed by technology substitution for energy in the 1970s leading to the world's highest energy efficiency improvement and broad advances in manufacturing technology level in the 1980s. Although Japan's dynamism shifted to the opposite in the 1990s, resulting in a lost decade due to a systems conflict between indigenous institutional systems and a new paradigm in an information society, a swell of reactivation emerged in the early 2000s.

This can largely be attributed to hybrid management fusing the "East" (indigenous strength) and the "West" (lessons from an IT driven new economy) typically observed in mobile driven innovation. Noteworthy success in such hybrid management can be seen in Canon, which effectively utilizes its indigenous strength ("East") in assimilating external technology ("West") while preserving its own organization by not depending on M&A. M&A reacted to deteriorate indigenous organization which does not necessarily adapt to exotic systems in Japan. Since Japan traditionally depends on its indigenous business model such as lifetime employment and seniority systems unique to individual firms, majority of M&A, particularly with foreign firms does not assimilate in such system. However, as a consequence of the fusing efforts, certain firms have shown noteworthy accomplishments in the synergy of M&A leading to a dramatic increase from 2004. The effect of fusing "East" and "West" validates the significance of global coevolution that corresponds to SIMOT's aim to elucidate, conceptualize, and operationalize Japan's explicit coevolutionary dynamism. This then enables them to accrue as global assets, thereon establishing a new innovative science, the "Science of Institutional MOT," that will enable any country with different institutional systems to effectively utilize its MOT.

10.2 Japan's Indigenous Explicit Function

10.2.1 Japan's Development Trajectory: Historical Perspectives

Japan indigenously incorporates an explicit function leading to technology substitution for scarce resources, which enabled firms to achieve conspicuous performance in a virtuous cycle between innovation and rapid economic growth in the 1960s. This was followed by technology substitution for energy in the 1970s leading to the

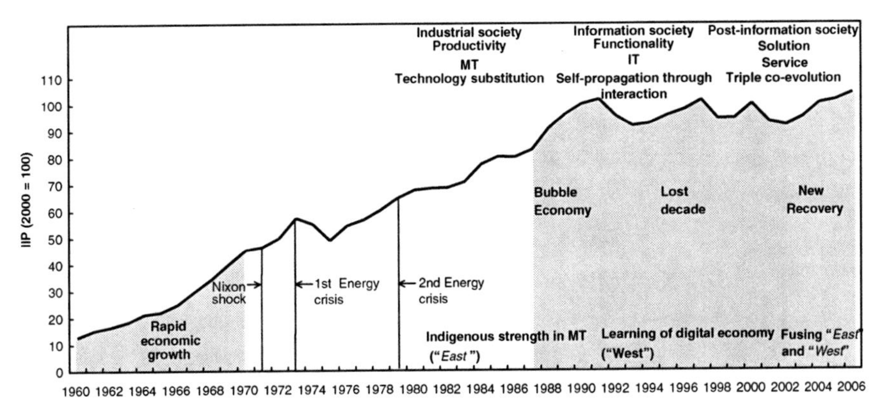

Fig. 10.1 Japan's development trajectory (1960–2006)
Source: Ministry of Economy, Trade and Industry, Index of Industrial Production (annual issue) [3]

world's highest energy efficiency and advances in manufacturing technology (MT) called high-technology miracles in the 1980s as demonstrated in Watanabe [5]. However, Japan's institutional characteristics of nonliquidity of work force resulted in institutional nonelasticity against paradigm shift to an information society.

New recovery in the early 2000s can be attributed to the fusion of indigenous strength in MT ("East") developed in an industrial society and effects of the cumulative learning of digital economy ("West") in an information society. This can be attributed also to Japan's explicit function in learning and fusing. Figure 10.1 demonstrates such a trajectory.

10.2.2 Japan Indigenous Institutional Systems for Innovation

Emerged innovation improves institutional systems such as economic, social/cultural, and natural environments, which in turn induces further innovation (coevolution) as demonstrated in Fig. 10.2. Innovation generation cycle leads to emerging innovations to market by means of effective utilization of resources in innovation. This inducement may stagnate if institutional systems cannot adapt to evolving innovation (disengagement).

10.2.3 Inducement of Innovation: Overcoming the Growth Constraints

Japan's development paths after WWII correspond to overcoming the growth constraints in each respective period: (a) 1950s: capital; (b) 1960s: labor; (c) 1970s:

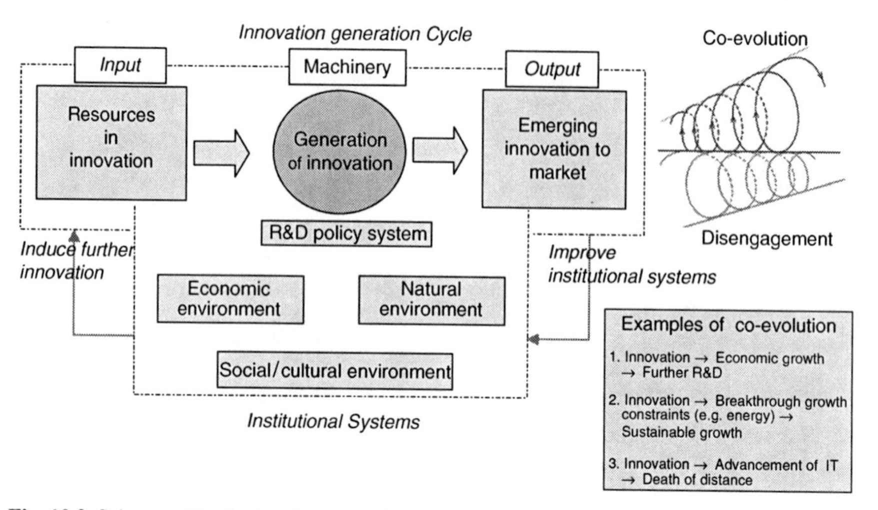

Fig. 10.2 Scheme of institutional systems for innovation

energy; (d) 1980s: qualified labor. Scarcity reflects prices of production factors. Every effort was focused on the productivity increase in scarce resources → technology substitution for scarce resources.

10.2.4 Technology Substitution for Constrained Production Factors

Supported by institutional systems for innovation, technology substitution for scarce resources functioned well in Japan typically in technology substituted for energy started from 1973 as clearly demonstrated in Fig. 10.3 [4]. This substitution not only enabled Japan to overcome energy constraints but also induced further innovation leading to constructing a coevolution between innovation and institutional systems. Thus Japan enjoyed high-technology miracle in the 1980s.

10.2.5 Conspicuous Energy Efficiency and World Top Level Manufacturing Technology

Japan accomplished the highest GDP growth as 3.97% p.a. in a decade after the second energy crisis in 1979. This can be attributed to its conspicuous energy efficiency enabled by technology substitution for energy as compared in Fig. 10.4. Consequently Japan demonstrates the world's highest energy efficiency as in Fig. 10.5. Technology substitution for scarce resources led Japan to demonstrate the world's top level manufacturing technology.

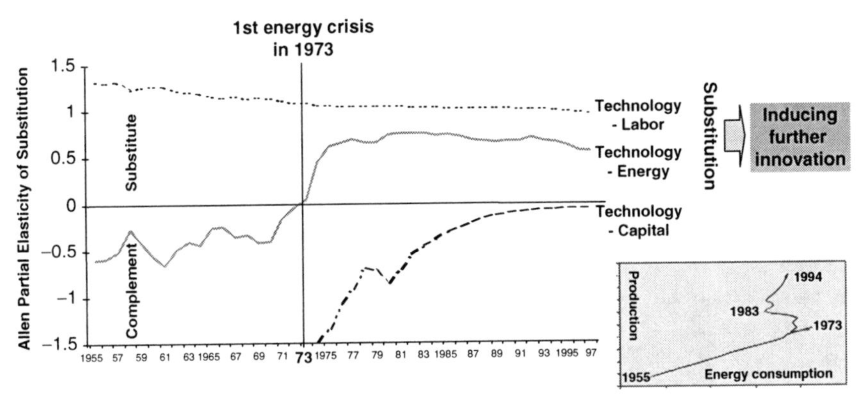

Fig. 10.3 Trends in substitution and complement among labor, capital, energy, and technology in the Japanese manufacturing industry (1955–1997) – Allen partial elasticity of substitution
Source: Watanabe (1999)

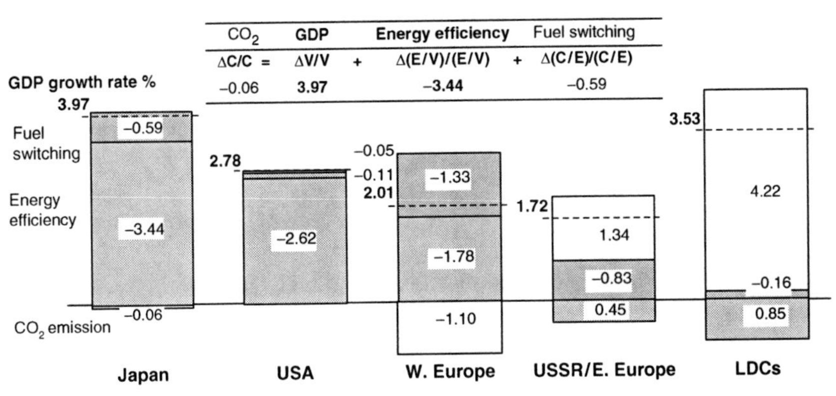

Fig. 10.4 Comparison of growth paths in major countries (1979–1988) – % p.a.
Source: Watanabe (1999) [4]

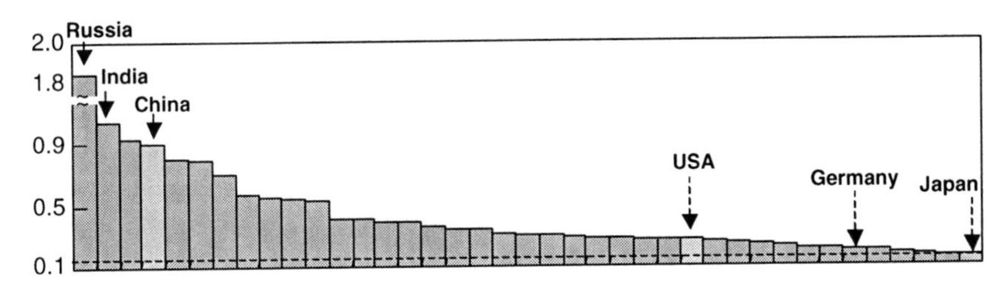

Fig. 10.5 Energy consumption per GDP in 40 Countries (2004)

10.3 Lost Decade and Reactivation

Although Japan's dynamism shifted to the opposite in the 1990s, resulting in a lost decade due to a systems conflict between indigenous institutional systems and a new paradigm in an information society, a swell of reactivation emerged in the early 2000s.

This can largely be attributed to hybrid management fusing the "East" (Japan's indigenous strength) and the "West" (lessons from an IT driven new economy) typically observed in mobile driven innovation.

10.3.1 Contrast Between Coevolution and Disengagement

10.3.1.1 Japan's Contrast Between Coevolution and Disengagement

Contrary to the high technology miracle in the 1980s, Japan experienced a long-lasting economic stagnation in the 1990s. This contrast can be attributed to a co-evolution between innovation and institutions in an industrial society (technology substitution for materialized production factors) and its disengagement in an information society (de-materializing society).

10.3.1.2 Features Differences Between Manufacturing Technology and IT

Disengagement in an information society is due to a system conflict toward de-materializing society [7]. Japan's conspicuous technology substitution for scarce resources functioned well for materialized production factors. However, as paradigm shifts to an information society, its subsequent shift from manufacturing technology to IT led to de-materializing society. Organizational inertia in an industrial society impeded Japan's institutions corresponding to an information society as demonstrated in Fig. 10.6 [10].

10.3.1.3 System Conflict in An Information Society

System conflict led to an institutional less-elasticity in an information society resulting in a dramatic decrease in MPT (marginal productivity of technology). MPT decrease led to TFP decrease resulting in a decrease in innovation contribution to growth. Thus coevolution changed to disengagement in an information society Figs. 10.7–10.9 demonstrates these trends.

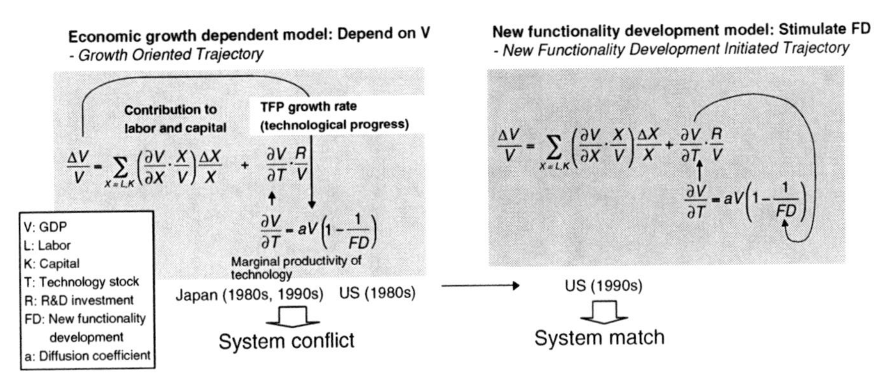

Fig. 10.6 Scheme leading Japan to lose its institutional elasticity
Source: Watanabe et al. (2003)

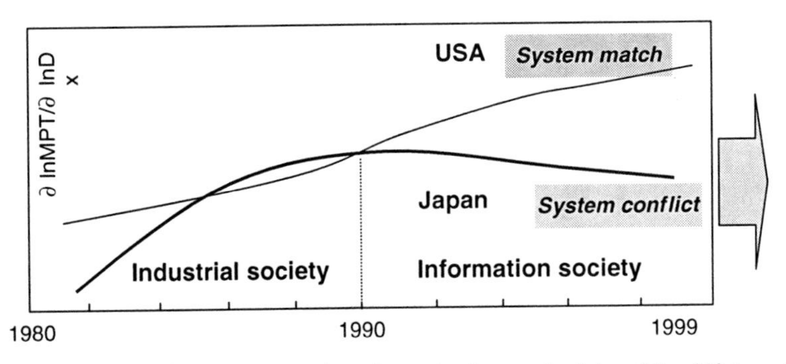

Fig. 10.7 Institutional elasticity of manufacturing technology – elasticity of the shift to an information society to marginal productivity of technology (1980–1999) – index: 1990 = 100

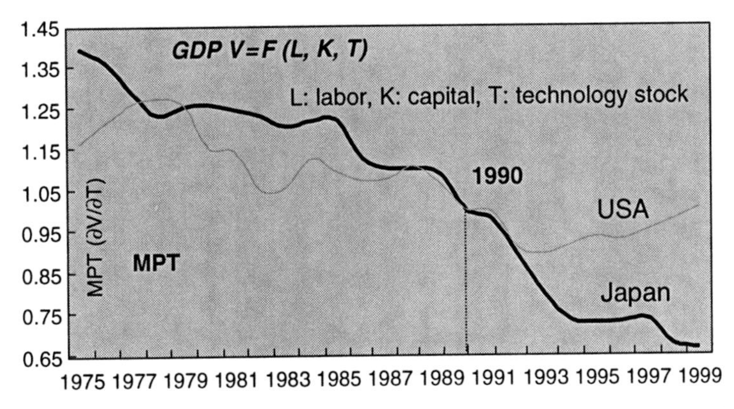

Fig. 10.8 Marginal productivity of manufacturing technology (1975–1999) – index: 1990 = 1

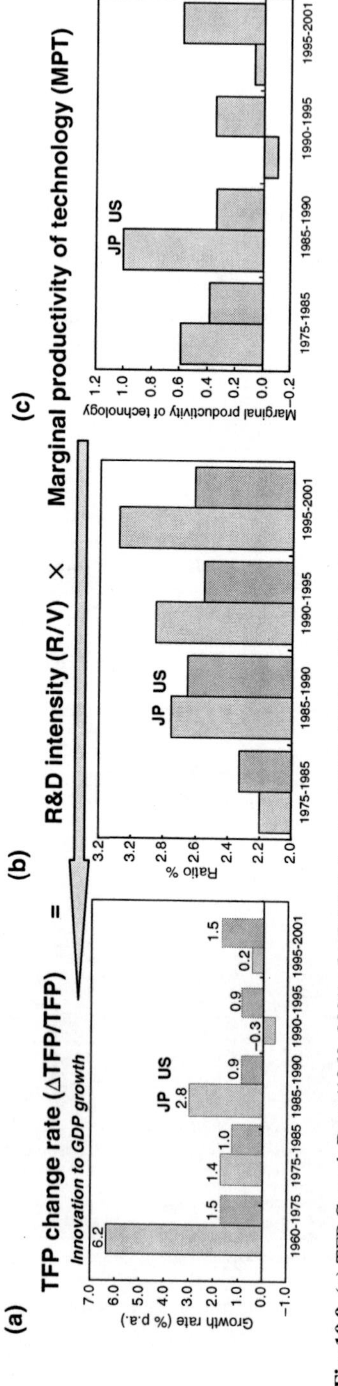

Fig. 10.9 (**a**) TFP Growth Rate (1960–2001). (**b**) R&D Intensity (1975–2001). (**c**) Marginal productivity of technology (1960–2001)

10.3.2 *Reactivation of Coevolutionary Dynamism*

While Japan's explicit coevolutionary dynamism shifted to opposite in the 1990s, surge of the reactivation was observed from the beginning of the 1990s typically in the innovation triggered by mobile phone driven innovation.

10.3.2.1 Surge of the Reactivation of Coevolutionary Dynamism

Shifting trajectory from MT (manf. technology) to IT (inform. technology). Mobile phone-driven innovation triggered a surge in coevolution. Dramatic increase in mobile phone exceeded fixed phones in 1998. The i-mode service from Feb. 1999 accelerated IP mobile diffusion [1]. Intensive interaction with institutions increased learning coefficient as demonstrated in Figs. 10.10–10.12.

10.3.2.2 Learning Enhances Functionality: Demand Structure

Increased learning coefficient enhances functionality development, leading to an increase in MPT and subsequent TFP growth.

10.3.2.3 Customized Operator–Vendor Structure: Supply Structure

In order to respond to such high functionality requirement, Japan's mobile handsets are focused on "customized types (93%) for high functionality classes (42% high-end ratio while world average is 7.8%)" [2].

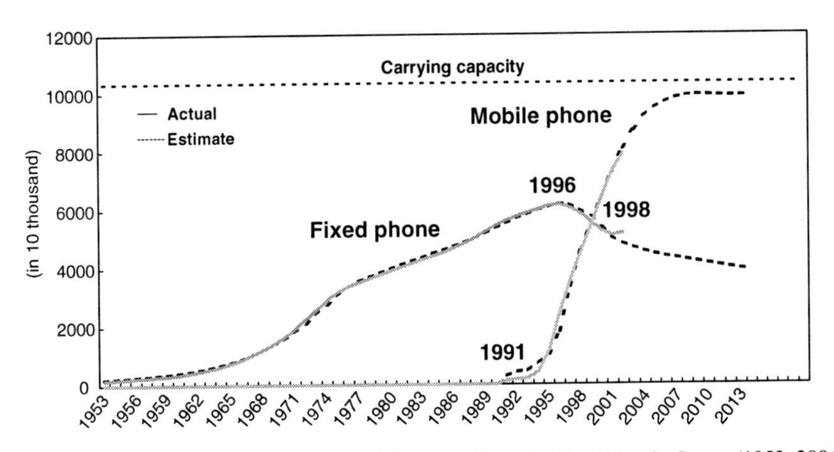

Fig. 10.10 Trend in the substitution of mobile subscribers for fixed line in Japan (1953–2004: actual; 2005–2015: estimate)

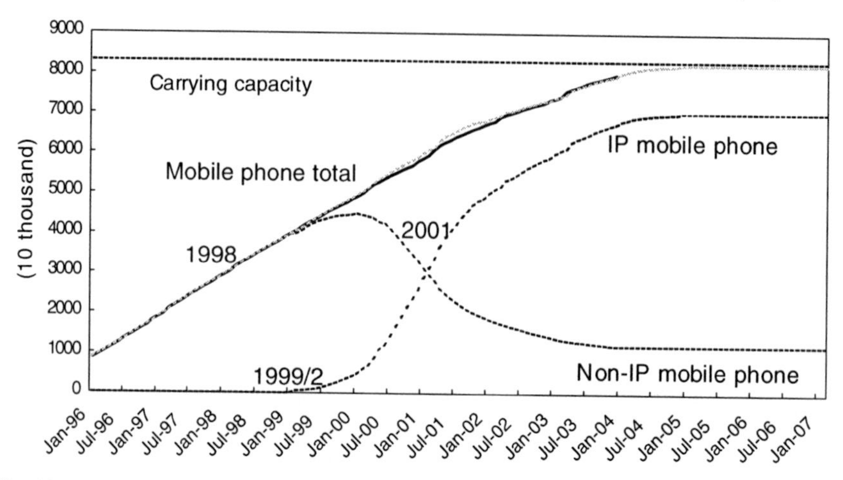

Fig. 10.11 Trend in mobile phone diffusion in Japan (Jan 1996–Dec 2003: actual, and Jan 2004–Jan 2007: estimate)
Source: Chen and Watanabe (2005) by means of choice–based substitution diffusion model

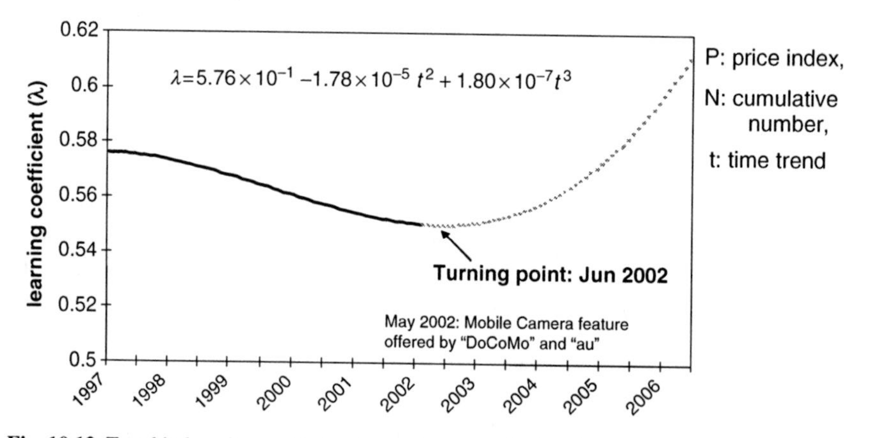

Fig. 10.12 Trend in learning coefficient in mobile phone (1997–2006)
Source: Chen and Watanabe (2006) [1]
λ = Dynamic learning coefficient, which indicates trend in the extent of learning effects incorporated in mobile phone. Figure 10.11 demonstrates substituting dynamism between mobile phone without and with IP function. Trend in IP mobile phone substitution for non-IP mobile phone demonstrates that IP exceeded non-IP in 2001

10.3.2.4 Coevolutionary Dynamism Between Operators and Vendors

Such requirements can be satisfied by tight operator–vendor relationship leading to co-evolution between them. This co-evolution is Japan's identical closed system (e.g. automobile and its parts, printers and PCs).

10.3.2.5 Coevolutionary Dynamism Leading to New Innovation

Dual coevolution in Japan's mobile phone development. Thus, dual coevolution through (a) market learning, and (b) operators–vendors interaction has been constructed in Japan's mobile phone development.

10.3.2.6 Reactivation of the Coevolutionary Dynamism: Mobile Phone Driven Innovation

This dual coevolution has induced Japan's mobile phone driven innovation that emerged in the beginning of the 2000s. While this innovation has stimulated the reactivation of Japan's indigenous coevolution between innovation and institutions, there remains the limit of the global deployment as it depends on closed suppliers chain as demonstrated in Fig. 10.13.

Mobile phone driven innovation stimulated activation of economy. Firms associated with mobile phones and their allied business activities (MPF) demonstrate a conspicuous IPO accomplishment (0.4 in years for IPO) and subsequent rapid sales increase (60% higher).

10.4 Hybrid Management: Fuses East and West

Japan is emerging from years of sluggish growth. Its firms appear to have produced something. Management method incorporates lessons from US firms while preserving the practices that once made Japanese firms famous. Figure 10.14 demonstrates such a concept.

10.4.1 Japan's Indigenous Potential in Fusing: Learning and Assimilation

10.4.1.1 System of Japan's Unique Learning Function

Japan indigenously incorporates explicit fusing potential through intensive cumulative learning efforts with its unique function [6, 8].

10.4.1.2 Learning and Assimilation

Based on this unique function, Japan's system of MOT has incorporated the following coevolutionary development cycle.

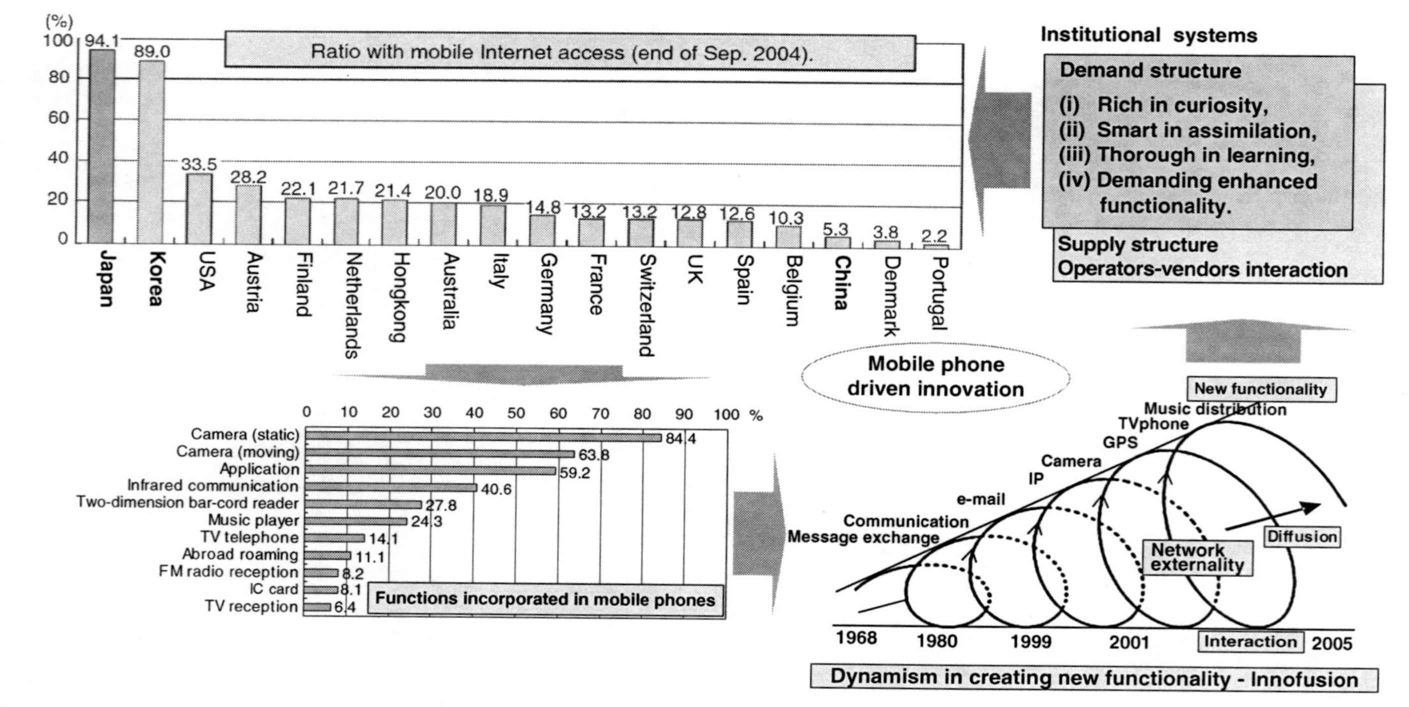

Fig. 10.13 Coevolution of mobile phone driven innovation

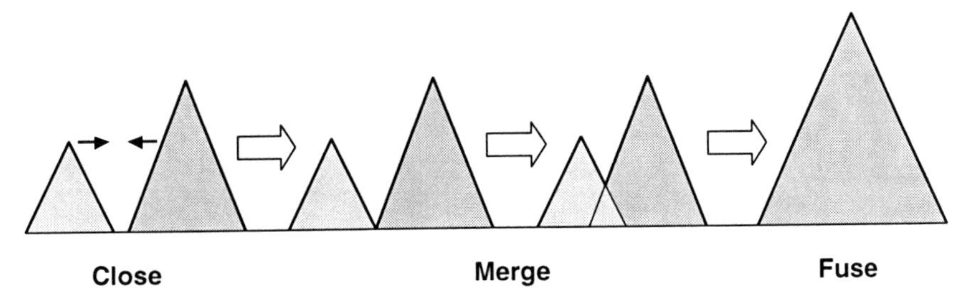

Fig. 10.14 Scheme of fusion

10.4.1.3 Conspicuous Assimilation Capacity

Consequently Japan accomplished high-performance learning in the 1980s as demonstrated by conspicuous assimilation capacity in 1993.

10.4.1.4 Dramatic Decline in Learning Effects in the 1990s

However, such learning efforts changed to negative in the 1990s due to

(a) X-inefficiency in a puffing-up with a success in the 1980s.
(b) Organizational inertia impeding flexible adaptation to a new paradigm of an information society and a mature economy.

10.4.1.5 Learning Efforts of Leading High-Technology Firms

Certain high-technology firms have maintained fusing efforts by endeavoring intensive learning as demonstrated in Fig. 10.15. Canon can be one typical example of the learned system LSI, SCM and cell production from external market.

10.4.2 Swell of Japan's Institutional MOT Toward a Post-Information Society

As a consequence of hybrid management fusing "East" and "West", Japan's indigenous MOT is again responding to a coevolutionary dynamism between innovation and institutional systems corresponding to a ubiquitous economy.

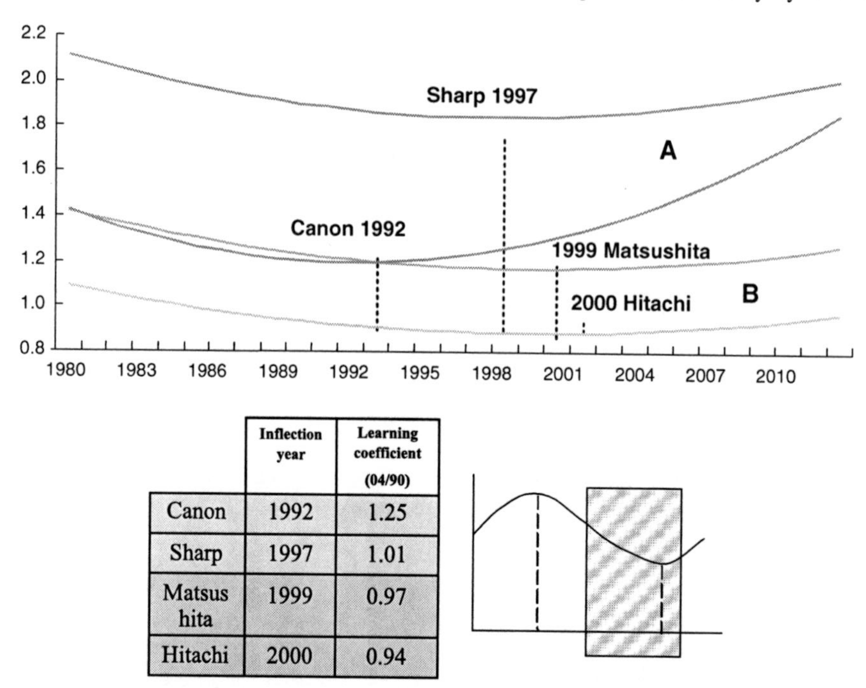

	Inflection year	Learning coefficient (04/90)
Canon	1992	1.25
Sharp	1997	1.01
Matsus hita	1999	0.97
Hitachi	2000	0.94

Fig. 10.15 Learning coefficients in four electrical machinery firms (1980–2003)
[a] Correlation between price of technology and governing factors of learning coefficients in Japan's leading machinery firms (1980–2003)

10.5 Canon's Success in Hybrid Management

Noteworthy success in such hybrid management can be seen in the performance of Canon, which effectively utilizes its indigenous strength ("East") developed through intratechnology spillover based on technological diversification strategy in assimilating external technology ("West") through "in vitro fertilization" (IVT) and "coopetition" (cooperation and competition between printer and PC producers) while preserving its own organization by not depending on mergers and acquisitions (M&A).

10.5.1 Canon's Conspicuous Performance

Canon accomplished conspicuous performance as demonstrated by extremely high level of OIS as in Fig. 10.16 [9, 11].

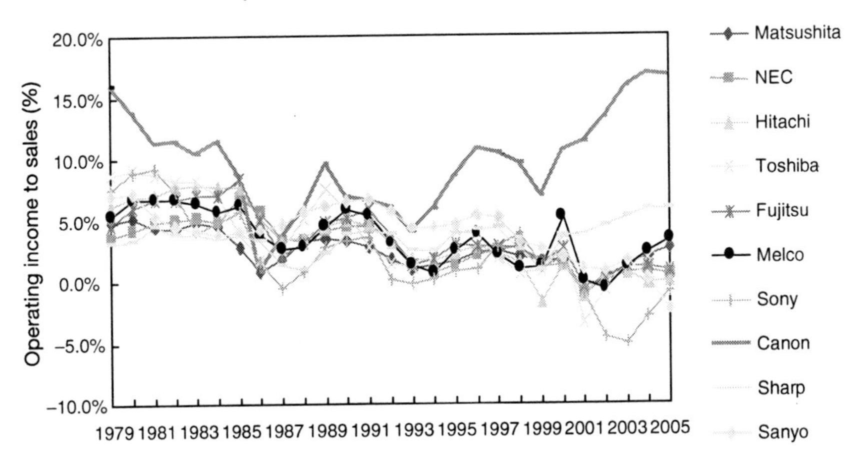

Fig. 10.16 Trend in OIS (1980–2005)

10.5.2 Functionality Development as a Source of High-Performance

(1) Functionality development,
(2) Prolongation effort to functionality development, and
(3) Successive innovation for sustainable functionality development

Canon's successive innovation from LLBP to LBP/BJ[1] contributed to enhance functionality development as demonstrated in Fig. 10.17.

The effect of fusing "East" and "West" with the synergy effects of M&A validates the significance of global coevolution for sustainable development that corresponds to this COE Center's aim to elucidate, conceptualize, and operationalize Japan's explicit coevolutionary dynamism between innovation and institutional systems.

This then enables them to accrue as global assets, thereon establishing a new innovative science, the "Science of Institutional MOT (SIMOT)," that will enable any country with different institutional systems to effectively utilize its MOT.

Canon accomplished two factors learning not only by its own technology stock but also by inspiring competitors. This is called coopetition. Thus, Canon's hybrid management consists of

(a) Market stimulation by providing attractive innovation (e.g. digital camera),
(b) Institutional technology spillover activating self-propagation,
(c) In vitro fertilization leveraging vendors innovation,
(d) Domestication by interfirm technology spillover through coopetition, and
(e) Intrafirm technology spillover emerging innovation efficiently as demonstrated in Fig. 10.18.

[1] LLBP, large scale laser beam printers; LBP: laser beam printers; BJ, bubble jet printers.

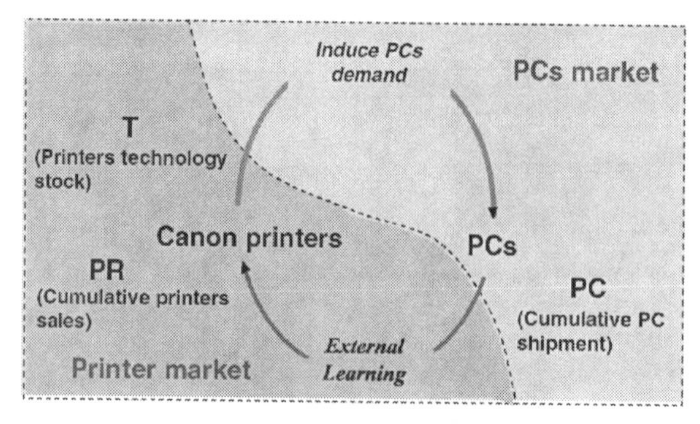

Fig. 10.17 Virtuous cycle between Canon printers and PCs (1986–1998)

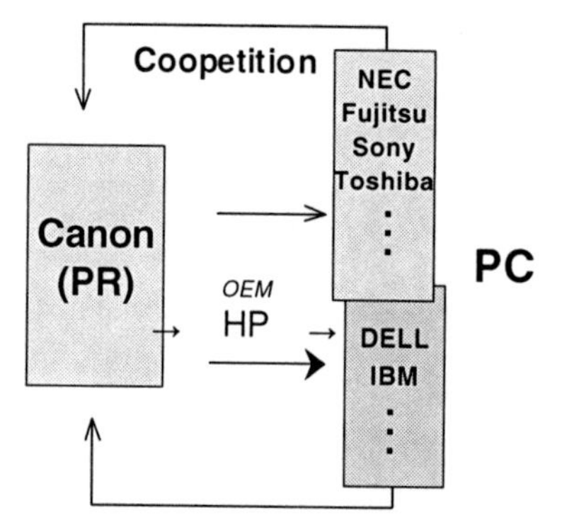

Canon's hybrid management enables coevolutionary domestication that can be attributed to two factors learning (TFLs) through coopetition based on fusing efforts.

10.5.3 Fusing Option

Canon's conspicuous performance can largely be attributed to intensive fusing efforts. It assimilates technology through "IVF and coopetition" while preserving its own organization by not depending on M&A.

M&A reacted to deteriorate indigenous organization and performance in Japan's manufacturing industry as they are the crystal of transgenerational improvement in the synergy effects of M&A by improving marginal productivity of technology as demonstrated in Table 10.1. While M&A reacted to decrease in Marginal Productivity of Technology (MPT) in leading high-tech firms such as NEC, Hitachi, Toshiba, and MELCO until the end of the 1990s, noteworthy opposite trend as increase in MPT has been observed from 2000 to 2002.

Confidence of such accomplishments has encouraged Japan's manufacturing firms to challenge M&A leading to a dramatic increase from 2004 as demonstrated in Fig. 10.19.

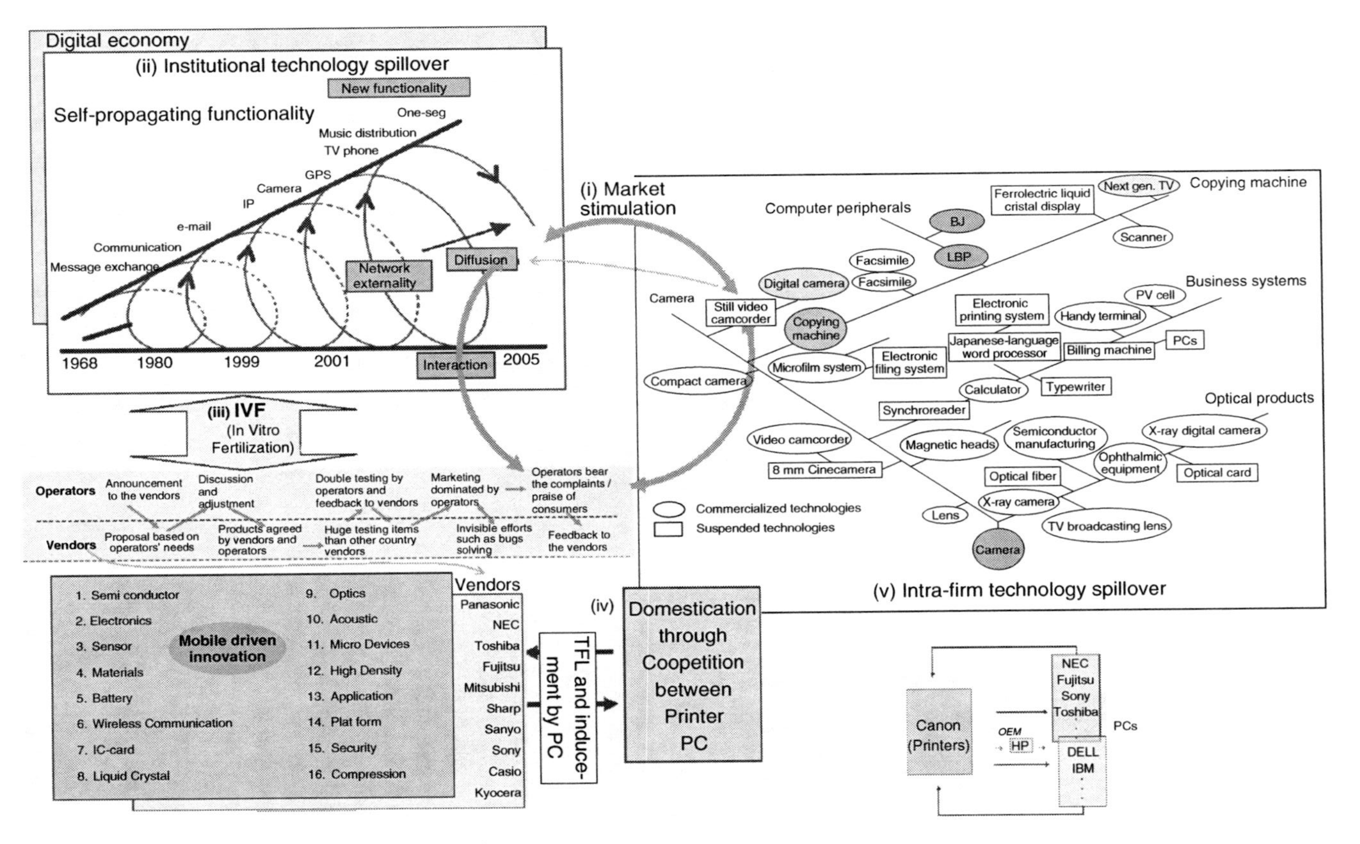

Fig. 10.18 Scheme of Canon's coevolutionary domestication

Table 10.1 Comparison of marginal productivity of technology in fusing firms by types of fusing

	1994–1996	1997–1999	2000–2002	2003–2005
Not by M&A (Canon, Sharp)	2.65	2.72	2.84	3.58
By M&A (NEC, Hitachi, Toshiba, MELCO)	2.50	2.30	2.45	2.94

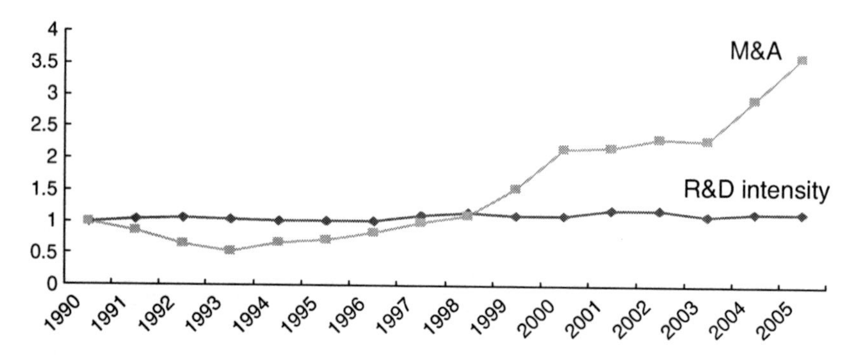

Fig. 10.19 Trend in M&A and R&D intensity in Japan's manufacturing industry (1990–2005) – index: 1990 = 1

10.5.4 Global Co-evolution for Sustainable Development

The coevolutionary dynamism between innovation and institutional systems is decisive for an innovation-driven economy. Rise and fall of the Japanese economy over the last three decades can be attributed to the consequence of the coevolution and disengagement between innovation and institutional systems as demonstrated in Fig. 10.20. Noteworthy surge in new innovation in leading edge activities in certain high-tech firms can be attributed to the fusion between indigenous strength developed in an industrial society ("East") and the effects of learning in an information society ("West").

This surge suggests a possibility of reactivation of the system of MOT leading to revitalizing Japan's economy. This can be enabled by constructing a virtuous cycle with vitalized world economy. All validates the significance of global coevolution for sustainable development as demonstrated in Fig. 10.20, which shows that sustainable development in the partner countries is indispensable for Japan's own development, which in turn induces further sustainable development in partner countries [12].

10.6 Conclusion

(a) Coevolutionary dynamism between innovation and institutional systems is decisive for an innovation-driven economy which may stagnate if institutional systems cannot adapt to innovations, and Japan's economy in the 1990s is one example.

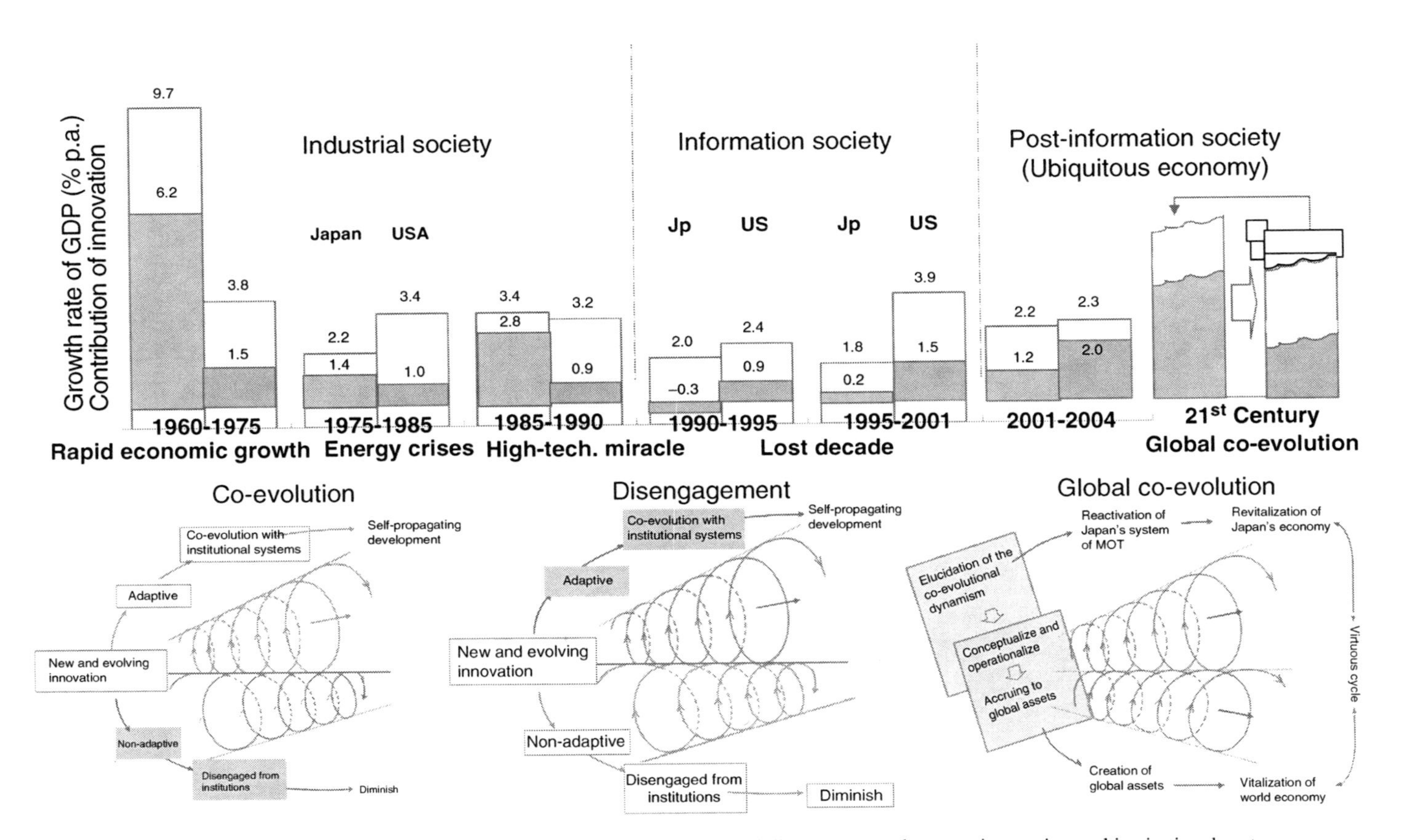

Fig. 10.20 Rise and fall of the Japanese economy as a consequence of coevolution and disengagement between innovation and institutional systems

(b) Japan indigenously incorporates an explicit function which induces coevolutionary dynamism enabling it to achieve conspicuous performance in a virtuous cycle between innovation and rapid economic growth in the 1960s followed by technology substitution for energy in the 1970s leading to the world's highest energy efficiency improvement and broad advances in manufacturing technology level in the 1980s.
(c) Although Japan's dynamism shifted to the opposite in the 1990s as disengaging the virtuous coevolutionship between innovation and advancement of its institutional systems, resulting in a lost decade due to a systems conflict between indigenous institutional systems and a new paradigm in an information society, a swell of reactivation emerged in the early 2000s.
(d) This can largely be attributed to hybrid management fusing the "East" (indigenous strength) and the "West" (lessons from an IT driven new economy) typically observed in mobile driven innovation.
(e) Noteworthy success in such hybrid management can be seen in Canon, which effectively utilizes its indigenous strength ("East") in assimilating external technology ("West") while preserving its own organization by not depending on M&A. M&A reacted to deteriorate indigenous organization which does not necessarily adapt to exotic systems in Japan due to Japan's unique traditional systems as lifetime employment and seniority system. However, as a consequence of the fusing efforts, certain firms have shown noteworthy accomplishments in the synergy of M&A leading to a dramatic increase from 2004.
(f) The effect of fusing "East" and "West" validates the significance of global coevolution that corresponds to SIMOT's aim to elucidate, conceptualize, and operationalize Japan's explicit coevolutionary dynamism. This then enables them to accrue as global assets, thereon establishing a new innovative science, the "Science of Institutional MOT," that will enable any country with different institutional systems to effectively utilize its MOT.

References

1. C. Chen and C. Watanabe, Diffusion, substitution and competition dynamism inside the ICT market: a case of Japan, Technology Forecasting and Social Change 73, No. 6 (2006) 731–759
2. C. Chen, C. Watanabe and C. Griffy-Brown, The co-evolution process of technological innovation: an empirical study of mobile phone vendors and telecommunication service operators in Japan, Technology in Society 29, No. 1 (2007) 1–22
3. Ministry of Economy, Trade and Industry, Index of Industrial Production (annual issue)
4. C. Watanabe, choosing energy technologies – the Japanese approach, in IEA (eds.), Comparing Energy Technologies (OECD/IEA, Paris, 1996) 105–138
5. C. Watanabe, Systems option for sustainable development, Research Policy 28, No. 7 (1999) 719–749
6. C. Watanabe, C. Griffy-Brown, B. Zhu and A. Nagamatsu, Inter-firm technology spillover and the creatation of a 'virtuous cycle' between R&D, market growth, and price reduction: the case of photovoltaic power generation development in Japan, in A. Gruebler, N. Nakicenovic and W.D. Nordhaus (eds.), Technological Change and the Environment (Resources for the Future (RFF) Press, Washington, DC, 2002) (ISBN 1-891853-46-5) 127–159

7. C. Watanabe, R. Kondo, N. Ouchi, H. Wei and C. Griffy-Brown, Institutional elasticity as a significant driver of IT functionality development, Technological Forecasting and Social Change 71, No. 7 (2004) 723–750
8. C. Watanabe and B. Asgari, Impacts of functionality development on the dynamism between learning and diffusion of technology, Technovation 24, No. 8 (2004) 651–664
9. C. Watanabe, J.Y. Hur and S.Y. Lei, Converging trend of innovation efforts in high technology firms under paradigm shift: a case of Japan's electrical machinery, OMEGA 34, No. 2 (2006) 178–188
10. C. Watanabe, H. Takahashi, Y. Tou and K.L. Shum, Inter-fields technology spillovers leveraging co-evolution between core technologies and their application to new fields: service-oriented manufacturing toward a ubiquitous society, Journal of Services Research 6, No. 2 (2006) 7–24
11. C. Watanabe and S. Lei, The role of techno-countervailing power in inducing the development and dissemination of new functionality: an analysis of Canon printers and Japan's personal computers, International Journal of Technology Management 44, Nos. 1/2 (2008) 205–233
12. W. Zhao, Co-evolution between software innovation and institutions: elucidation of co-evolutionary dynamism between Japan and China through outsourcing (Master Thesis, Tokyo Institute of Technology, Tokyo, 2006)

Chapter 11
Conclusion

Abstract This chapter summarizes the noteworthy findings obtained through nine dimensional analyses introduced in the preceding nine chapters and subsequent policy implications suggested in each respective chapters suggestive to firm entrepreneurial strategies toward post information society. Given the current global economic stagnation, these suggestions should be addressed more to reactivating the hybrid management partners so as maintain the effects of hybrid management of technology.

11.1 Formation of IT Features Through Interaction with Institutional Systems: *Empirical Evidence of Unique Epidemic Behavior*

In light of the understanding that effective utilization of potential benefits of dramatic advancement of IT in an information society will differ greatly depending on the nation, particularly on their institutional elasticity and that this can be attributed to the specific features of IT which performs its function in connection with institutional systems, this paper attempts to derive specific IT features by focusing on the unique diffusion process, in other words epidemic behavior of IT.

An empirical analysis on the diffusion process of innovative goods in Japan was conducted taking refrigerators, color TV sets and cellular telephones which represent innovative goods centered by manufacturing technology and IT. On the basis of the comparative analysis of epidemic behavior between IT and other technologies using the simple logistic growth function, bi-logistic growth function and logistic growth function within a dynamic carrying capacity, it was demonstrated that the specific features of IT are formed through dynamic interaction with an institutional system. In addition, certain specific features of IT, characterized during the course of the interaction process and conspicuous in its unique epidemic behavior, were identified including disseminative, interactive, coevolutional as well as extremely invisible and global than technology in general. Furthermore, a mechanism of IT's contribution to increasing returns to scale was identified.

C. Watanabe, *Managing Innovation in Japan: The Role Institutions Play in Helping or Hindering how Companies Develop Technology*, DOI: 10.1007/978-3-540-89272-4_11,

These analyses provided us significant insight that Japan's industrial and management system based on nonstylized management system unique to individual firms/organizations functioned well for innovation and diffusion of manufacturing technologies which supported industrial society. Furthermore, it has become evident that such nonstylized management system does not function well for innovation and diffusion of IT features, which are formed through dynamic interaction with an institutional system to which stylized management system is indispensable.

All these findings remind us the significance of the role of institutional elasticity in making full utilization of potential benefit of IT and also of the urgency of remediation of Japan's lost institutional elasticity. Thus, system functions supportive to complement for remediation of the institutional elasticity with distinct features of IT would be crucial.

11.2 Institutional Elasticity as a Significant Driver of IT Functionality Development

The effective utilization of IT in an information society will differ greatly depending on structural differences in different countries. In particular, the elasticity or flexibility of institutions in the national innovation system is critical. Furthermore, the specific functionality of IT is developed through interactions with flexible institutional systems. This paper attempts to show this process by focusing on the unique diffusion process, or epidemic behavior of IT.

First, through the mathematical analysis of the diffusion process of innovative goods, the mechanism in creating a new carrying capacity in the process of IT diffusion was conceptualized. On the basis of this framework, an empirical analysis of Japan's manufacturing industry using a new approach to decomposing the factors of TFP was developed. This postulated that IT creates new demand in its development process and specific functionality is formed in this interactive system. Second, an empirical analysis of the diffusion process of innovative goods in Japan was conducted using refrigerators, fixed telephones, Japanese word processors, color TV sets, personal computers and cellular telephones which represent innovative goods across a spectrum from monofunction to multifunction products. Through a comparative analysis of epidemic behavior between these technologies using a logistic growth function within a dynamic carrying capacity, it was demonstrated that innovative goods with higher IT density matched the logistic growth function within a dynamic carrying capacity. These analyses demonstrated that the specific features of IT are formed through dynamic interaction with an institutional system. In addition, certain IT qualities enabling this self-propagating behavior were identified and discussed. These unique qualities, critical to enhancing the epidemic behavior observed, depend on institutional flexibility to "capture" the potential of IT. Furthermore, Japan's institutional system currently does not provide the flexibility required to fully exploit these features.

These findings remind us of the significance of the role of institutional elasticity in making full utilization of IT and also the urgency of re-creating institutional elasticity in Japan. While this paper briefly reviews this subject, substantial analysis is beyond the scope this work. Further analysis at the firm level, in terms of fundamental approaches to information and process systems as well as strategic management, is required to fully elucidate critical success factors in US and Japanese firms amidst the information society paradigm.

11.3 A Substitution Orbit Model of Competitive Innovations

In light of the increasing significance of timely introduction of emerging new technologies that substitute for existing technology for enhancing a nation's international competitiveness in a globalizing economy, focusing on Japan's transition from manufacturing technology to IT, analyzes substitution orbits of two competitive innovations were analyzed.

Prompted by a sophisticated balance of the coevolution process in an ecosystem, particularly of the complex interplay between competition and cooperation, an application of Lotka–Volterra equations that analyze these sophisticated balance in an ecosystem is conducted for reviewing the substitution orbits of Japan's monochrome to color TV system, fixed telephones to cellular telephones, and cellular telephones to mobile internet access service and analog to digital TV broadcasting.

On the basis of the comparative assessment, by using synthesized two-dimensional Lotka–Volterra equations for substitution and general logistic growth equation following findings are obtained:

(1) Lotka–Volterra equations for substitution are useful for identifying an optimal orbit of competitive innovations with complex orbit.
(2) It is particularly useful for assessing policy options from the viewpoint of effectiveness in controlling parameters for leading to the expected orbit.
(3) Under the global paradigm shift from an industrial society to an information society, given the government target to accomplish a rapid shift from traditional technology to new technology, particularly IT driven new technology within a limited period, shifting scenario should be accelerated in line with Lotka–Volterra orbit with optimal coefficients.
(4) In order to accomplish this orbit, every effort should be accelerated in removing factors separating the two orbits between logistic growth orbit and Lotka–Volterra orbit including:

    (a) A lack of information about new technologies expected to be substituted for traditional one,
    (b) Fear of substitution and a reluctances to pay the cost of switching from traditional to new technology, and
    (c) Barriers to prompt shift to new technology within producers, distributors and customers.

(5) In addition, in order to accelerate such shift with a higher pace of Lotka–Volterra orbit, with the understanding that IT's specific functionality is formed through dynamic interaction with institutional systems, efforts should be focused on maximizing IT's self-propagation behavior.

Considering that while hasty shift sometimes accomplishes nothing delayed shift can result in a loss of national competitiveness, identification of the optimal diffusion and substitution orbit is essential; thus, the foregoing approach is very useful in identifying policy options for the diffusion orbit of competitive innovations with complex interplay.

Further mathematical attempts aiming at broader application of possible orbit for substitution, together with empirical analyses taking "success stories" in smooth substitution and rapid diffusion such as mobile internet access service on factors, conditions and systems enabled them rapid substitution and diffusion to be expected to be undertaken.

11.4 Impacts of Functionality Development on Dynamism Between Learning and Diffusion of Technology

In light of the increasing significance of the systems approach in maximizing the effects of innovation by means of the effective utilization of the potential resources of innovation, this chapter undertook theoretical analysis of this subject focusing on a dynamism between learning and diffusion of technology. An empirical demonstration was also attempted taking Japan's PV development trajectory, which follows the similar trajectory of IT's functionality development, over the last quarter century.

On the basis of these analyses, dynamism between learning and diffusion of technology was elucidated, thereby the effects of functionality decrease on learning coefficient and consequent impacts on technology diffusion and its carrying capacity were identified.

Noteworthy findings include:

(1) On the basis of intensive empirical analyses and reviews of proceeding works, it was anticipated that the behavior of learning coefficient has close relevance with that of a logistic growth function within a dynamic carrying capacity. This coefficient was anticipated to increase as a consequence of cumulative learning effects and change to decreasing trend in the long run as functionality decreases.
(2) Such a dynamic convex behavior of learning coefficient was enumerated by an equation derived from a logistic growth function within a dynamic carrying capacity with an additional term reflecting functionality decrease in the long run. On the basis of an empirical analysis, by applying this equation in Japan's PV development trajectory over the last quarter century it was demonstrated that this equation reflected the learning coefficient of Japan's PV firms, thereby the significance of this equation was demonstrated. This dynamic coefficient function incorporating functionality decrease effects revealed that an estimate

without considering functionality decrease effects leads to higher estimate than that estimated by reflecting functionality decrease effects.

(3) Synchronizing this equation in a logistic growth function within a dynamic carrying capacity, an equation depicting diffusion trajectory of innovative goods incorporating functionality decrease effects was developed which demonstrates similar trajectory as actual one, thereby significance of this equation was demonstrated. A trajectory estimated by this equation demonstrates slightly lower diffusion trajectory than the trajectory estimated by a normal logistic growth function within a dynamic carrying capacity without considering functionality decrease effects. This was considered because of a "depression effect" as a consequent of functionality decrease.
(4) On the basis of this new logistic growth function within a dynamic carrying capacity incorporating functionality decrease effects, the impacts of this functionality decrease on the dynamic carrying capacity was analyzed. The analysis identified that this impact is not so significant in short term, but significant in the long run in stagnating carrying capacity as time passes by was revealed. In addition, it was identified that the decrease in this carrying capacity accelerates obsolescence of technology. This significant impact was identified to lead a vicious cycle between stagnating carrying capacity and diffusion trajectory.

Important suggestions, supportive to the nation's technology policy and firm's R&D strategy in light of the maximum utilization of the potential resources for innovation under a long-lasting economic stagnation while facing a new paradigm initiated by an information society, can be focused on the following points:

(1) Systems restructuring is indispensable for shifting a vicious cycle between stagnation of diffusion and carrying capacity. Given the IT's self-propagating nature formation process in which interaction with institutions plays a significant role, activation of interaction with institutional systems plays a significant role to this restructuring.
(2) In this consequence, a way to lead a positive dynamic interaction between learning, diffusion and spillovers of technology depends on institutional elasticity for activating interaction with institutional systems.

Given that the state of institutional systems constructs a virtuous cycle between techno-economic development of the nation, shifting current vicious cycle to a virtuous cycle would be crucial.

Points of further works are summarized as follows:

(1) Further elaboration of the relationship among the state of institutional system, more specifically institutional elasticity, the state of innovation and diffusion and trend in functionality.
(2) International comparison of the institutional elasticity and its effect on innovation and diffusion of technology.
(3) Demonstration of the significance of institutional elasticity and its contribution in maximizing the effects of policy.

11.5 Diffusion, Substitution and Competition Dynamism Inside the ICT Market: *A Case of Japan*

In light of the significance of the reconstruction of Japan's vicious cycle between nonelastic institutions and insufficient utilization of the potential benefits of ICT, this chapter analyzed diffusion, substitution and competition dynamism inside Japan's ICT market.

Prompted by the hypothetical view that recent advances in its IP mobile service deployment such as NTT DoCoMo's i-mode service can be attributed to a co-evolutionary dynamism between diffusion, substitution and competition inside its ICT market, an empirical analysis of the mechanism co-evolving this dynamism in telephony, mobile telephony and internet access markets was attempted by utilizing four types of diffusion models identical to respective diffusion, substitution and competition dynamics.

Noteworthy findings in each respective market include:

(1) Telephony market

The simple logistic model demonstrate that the potential of fixed line phone market in Japan is about 62 million, lower than that of the mobile phone since fixed line phone is usually shared by people who live together such as family members while mobile phone is always possessed by an individual.

(a) However, the subscription of fixed line phone has started to decline after reaching the peak in 1996. It is considered that the complementary feature between mobile and fixed line telephone is the reason for such decrease in a matured fixed phone line market.

(b) By taking mobile and fixed line telephony as two generations of telephony with substitutability and analyzing with the choice-based substitution diffusion model, it is demonstrated that the potential of the overall telephony can be about 100 million, consisting of double subscribers who have signed up to both mobile phone and fixed line phone.

(c) Consumers' utility to adopt fixed line phone increase as time goes by before mobile phone is introduced, but deceases with time after mobile phone appeared in the market. The cost of monthly subscription fee of mobile phone is a negative factor for users to adopt or switch to mobile phone; moreover, the double-subscribers reveal the highest cost elasticity and nonsubscribers seem to consider cost less than those who have already subscribed to either one service. The result can be attributed to the fact that people who have not yet adopted any telephony service have higher need to subscribe than those who just consider to switch.

(d) Moreover, even after mobile phone is introduced into the market, existing subscribers of fixed line phone still tend to continue such subscription as long as the price of mobile phone does not drop too much.

(e) In both mobile and fixed line telephony cases, it is observed that voice-only service is occupying less and less share of communication market while data-and-voice service is growing. With only mobile and fixed line telephony taken into consideration, the substitution effect lead the subscription of fixed line phone to decrease; however, including the impact by the internet access demand, the fixed line phone demand will decrease less than analyzed with this model.

(2) Mobile telephony market
Both simple logistic model and bi-logistic model demonstrate that the potential of mobile phone market in Japan is between 83 and 85 million, with the speed that half of the potential can be reached within about 13 years.

(a) While the diffusion process is divided into two impulses, the bi-logistic model demonstrates that the potential of market increased from 50 million to more than 83 million after the IP mobile phone was introduced into the market.
(b) However, with the choice-based substitution diffusion model that takes the substitution of NonIP by IP mobile phone into consideration, the potential of NonIP mobile phone was originally about 56 million, but the penetration rate slowed down before reaching this level and users switching from NonIP to IP mobile phone service make the diffusion of IP mobile phone move faster than expected. The overall potential of mobile phone is still estimated as about 83 million with this model.
(c) The analysis with the choice-based substitution diffusion model demonstrates the factors that affect consumers' adoption intension. Consumers' utility of adopting NonIP mobile phone increases as time goes by before IP mobile phone is introduced. However, after IP mobile phone appeared in the market, consumers' utility of adopting Non-IP decreases with time because of obsolescence. Users' utility of keep using Non-IP mobile phone increases with time but more slowly than that of adopting or switching to IP mobile. As expected, consumers are more likely to adopt to IP mobile phone with reduced price and enriched functions.
(d) Finally, on the basis of the previous analysis, analysis with logistic growth within a dynamic carrying capacity demonstrates that considering the positive impact of enriched functions on ICT development, the potential of mobile phone can actually reach to about 94 million, which is higher than the estimates with previously implemented models.
(e) Observations of the newly gained market share shows that the relatively stable market might be changed in the near future. This is an evidence of more and more competitive environment that is a positive driving force of mobile phone diffusion.

(3) Internet access market

(a) The simple logistic model demonstrates that the potential of the internet access market in Japan is about 37 million, even lower than that of the fixed line phone. The reason is considered to be that internet access diffusion is limited by personal computer possession rate. Moreover, fixed line internet access can also be shared by people who live together such as family members just as in the case of fixed line phone.

(b) While the diffusion process is divided into two impulses, dial-up access and broadband access, the bi-logistic model demonstrates that the potential of mobile phone market in Japan increased by about 12 million contributed by the broadband internet access service. The impulse caused by the broadband internet access grows at a speed faster than the first impulse of dial-up access. Such a difference in diffusion speed is attributed to the always-connected feature and the high access speed of broadband that is more appealing to the internet users.

(c) Moreover, with the choice-based substitution diffusion model that takes the substitution of dial-up access by broadband internet access into consideration, the potential of internet access was originally only about 8 million, but after broadband access is introduced, both subscription of dial-up access and broadband internet access increased dramatically.

(d) The analysis with the choice-based substitution diffusion model demonstrates that both time and price affect consumers' adoption intension. Time also represents the degree of maturity of internet technology as well as the bandwidth, thus it is rational to have the result that consumers' utility increases with time. Consumers' utility of using broadband access increases faster with time than that of using the dial-up service.

(e) However, the case of the internet is different from that of the telephony because users' utility of adopting dial-up access increases with time, no matter before or after broadband access is introduced. This reason is considered because people still sign up to dial-up service while broadband has already been introduced into the market since they consider themselves as nonfrequent users of the internet. Therefore, they can still be satisfied with the limited speed offered by the dial-up access.

(f) The penetration rate of the dial-up access, which has started to slow down since 2002, is expected to decline rapidly during the coming years. Users switching from the dial-up access to broadband access make the diffusion of the broadband access move faster than expected. The overall potential of the internet access is estimated to be about 40 million with the choice-based substitution diffusion model.

(g) Competition is expected to be the most important driving force to fill up the gap between the broadband market development in Japan and Korea.

(4) Diffusion, substitution and competition dynamism inside the transitional market in telephony, mobile and internet.
All these findings obtained in the substitution dynamism in the markets of telephony (fixed line to mobile), mobile telephony (Non-IP mobile to IP mobile) and internet access (dial-up to broadband) demonstrate a noted co-evolutionary dynamism between diffusion, substitution and competition emerging inside the Japanese ICT market in transition. Key factors governing this dynamism are identified as ICT innovation, enriched functions, reduced price and competitive environment.

These findings provide important policy implications and can be summarized as follows:

Given that the foregoing dynamism can be the souce of Japan's noted advances in IP mobile servile deployment, despite a lack of institutional elasticity, systems approach in stimulating a co-evolution between ICT innovation, diffusion, substitution and competition with the special attention to the following policies would be essential:

(a) With IP mobile phone as the mainstream of telephony market, it is expected that everyone can have a ubiquitous information receiver. Being able to communicate with others via not only voice but also data will definitely increase the mobility and hence improve the efficiency of necessary communication. Deregulations to promote competition are furthering smoothly in most advanced countries, and the government should continue to make effort to keep such competitive environment rolling. Moreover, the possibility of mobile phone may be extended from only e-mail receiver to more practical internet terminal. Making mobile internet a required function for a convenient life will increase its penetration rate more efficiently.
(b) With the substitution by mobile phone, fixed line phone seems to be considered as a sunset industry. However, judging from the increasing demand for data-and-voice service, it is possible that the demand for fixed line phone will increase again if the demand for the internet is significant.
(c) Currently, the broadband access still depends on the fixed line phone infrastructure, so the government should take such impact into consideration when deciding on whether to continue expanding fixed line phone infrastructure development or not.

Future works can be summarized as follows:

In this research, the empirical diffusion and substitution analysis of ICT market is focused on Japan's market. However, given the significance of such analysis done, with data of more countries, similar analysis should be conducted for other countries.

(1) In the part of telephony market analysis, it s noted that the ratio of double-subscribers are so difficult to capture that we can only facilitate the result of annual investigation. However, with the subscription rate reported monthly, double-subscriber ratio should also be collected on a monthly base if possible.
(2) The personal computer market is expected to reveal similar characteristics of diffusion and substitution. However, unlike services that users have to subscribe to, personal computers are hardware that consumers buy and use. It is more difficult to keep track of the real possession rate of personal computers. Similar analysis should be conducted with sufficient data of personal computers, if possible.
(3) On the basis of the result of personal computers diffusion analysis, the correlation between personal computers penetration rate and the internet access penetration rate can be analyzed to study their interaction, including the restriction that personal computers possession rate might have on the internet access rate

and the contribution that the internet access might have on personal computer purchasing.

(4) Finally, the whole ICT market, including the service section and the hardware section, can be analyzed from a comprehensive level. It is expected that the ICT policies can be made from a more comprehensive view with such analysis conducted well.

11.6 The Co-evolution Process of Technological Innovation: *An Empirical Study of Mobile Phone Vendors and Telecommunication Service Operators in Japan*

In light of the Japan's unique supply structure between operators and venders that contributes to dramatic advancement of mobile phones, this chapter elucidated the specific structure of the mobile phone market in Japan as follows:

The first part of the analysis classified the existing handset models showing that high-end handsets occupy the largest share in Japan's market, and that the ratio is much higher than the worldwide average level.

The second part of this analysis demonstrated that most domestic handset vendors offer customized models to satisfy the specific demands of each service operator.

Consequently, global handset vendors have difficulty entering the marketplace by simply offering global models to the service operators.

Similarly, because of the unique institutional structure in Japan, particularly the mutual dependence of the handset vendors and service operators on each other for their success, the Japanese vendors/service operators also struggle in other markets.

Owing to Japan's social institution, consumers' strong consciousness toward high quality, and innovative functions driving both service operators and handset vendors' commitment to quality, a closed but high-standard relationship between handset vendors and service operators has been developed which works closely with consumer demand.

This mutually-dependent relationship pushes the virtuous cycle of technological innovation so that it works smoothly and efficiently.

The following are the particularly notable new findings based on this analysis:

(1) Japan has become a market dominated by high-end handsets since customers in Japan prefer high quality and new functionality. According to the learning curve, it is observed that the start of mobile phone camera is the most important time point that triggered the learning coefficient turned upward.

(2) Japan has also become a market dominated by order-made handset development between service operators and handset vendors whose close tie-up supports this demanding market. Both service operators and handset vendors are responsible for the quality of handset, and the service operators, not the mobile phone end-users, are actually the immediate customers of the handset vendors.

(3) The interaction of the demand side and supply side in the mobile phone market of Japan formed a coevolution mechanism. It enables the extraordinarily high achievements in terms of quality and level of technology in this unique marketplace. However, on the other hand, the marketplace is relatively closed and also prevents other global handset vendors from succeeding in Japan making it difficult for Japanese vendors/operators to succeed in other markets, either.

On the basis of these unique findings, there are important implications for policy makers inside and outside Japan who are interested in creating an environment that stimulates an innovative marketplace of similar attributes. The following are some of the major policy implications:

(1) The first lesson learned is that one or two groundbreaking but user-friendly function expansion is necessary to stimulate the demand of a saturated market. The mobile camera feature combines handset and digital camera, and it showed the market a new possibility for using the mobile phone device. However, too many minor functions might only cause the consumers to spend more time to adapt and should be avoided.
(2) Handset vendors in Japan rely on the service vendors very much. This close relationship of R&D, manufacturing and sales cooperation drives the progress of the mobile phone in Japan, but may not be applicable in other markets where the service operators and handset vendors have to face the market separately. Japanese vendors and service operators should learn to be more independent from each other and more flexible in order to adapt themselves to the local supply-side structure in other markets.
(3) Although Japanese handset vendors and service operators are good at satisfying the demanding consumers in Japan with high-quality products and services, this high-cost business would be impossible without the sufficient support of consumers. The positive coevolution mechanism in Japan builds on the foundation of a unique social institution. To succeed in other markets, Japanese businesses have to adapt their strategies to the attributes of each market and learn to balance the cost and quality requirement. Another possible way is to wait or educate the consumers until they also request services and products at the same level as Japanese consumers.

Through this research, sources of Japan's coevolutionary dynamism between mobile-phone-driven innovation and its unique institutional systems have been identified. While this dynamism induces the dramatic advancement of mobile phones by enhancing functionality in a self-propagating manner, it incorporates structural constraints in embarking in a global market where more flexible options of both supply and demand structure are required.

Given that Japan's sustainable development depends on its own economic coevolution with global sustainability, Japan's mobile-phone-driven innovation should be globally expansive. Thus, Japan's mobile phone industry should shift from a homogeneously integrated structure that works well, but only in Japan, to a heterogeneous structure which is able to adapt and thrive amidst other demand-side factors. While Japan's current system appears to be homogenous, even in the relatively high-standard market in Japan, there are still differences among major players with respect to their strategic positioning, technology standards decisions and global strategies. While NTT DoCoMo's focus is on maintaining domestic supremacy in the consumer market with 3G capabilities by targeting "discerning customers" who want high quality service, au KDDI seems to be positioning itself for broader applications. How such differences in strategy and positioning may affect their achievement in penetrating into other markets is still a critical question worthy offurther exploration. After certain markets reach a saturation point, how to accelerate coevolution process not only in the domestic market but also in other markets will determine the long-term success of the firms involved. Future work should focus on this critical dimension.

11.7 Technopreneurial Trajectory Leading to Bipolarization of Entrepreneurial Contour in Japan's Leading Firms

In light of a swell of the reactivation of Japan's coevolutional dynamism leading to a noticeable contrast in high-level mobile phone dependency and subsequent mobile-phone-driven innovation, this chapter attempted to elucidate this coevolutionary dynamism.

Prompted by the notable contrast observed in a rapid increase in mobile phone diffusion, a conspicuous growth in high-functional fine ceramics, and also bipolarization trajectory in leading high-technology firms, an elucidation of the coevolutionary dynamism between transformation of the characterization of technology and subsequent change in entrepreneurial features was focused leading to the following empirical analyses:

(1) Coevolutionary dynamism between the advancement of mobile phones with internet access service and institutional change,
(2) Elucidation of the black box of new functionality development and its mechanism by means of cross-product technology spillover dynamism in high-performance fine ceramics, and
(3) The transformation to technopreneurship in leading Japanese firms confronting the transition to a postinformation society.

On the basis of the above analyses, noteworthy findings were obtained including:

(1) Transformation of characterization of technology through the course of interaction with institutions should be taken into firm's technopreneurial strategies,
(2) Institutional elasticity (flexibility, adaptability, cooperative and openness to foreign ideas) is essential to correspond to such strategies,
(3) Mobile phone driven innovation induces a coevolutionary dynamism between innovation and institutions leading to leverage the reactivation of Japan indigenous coevolutionary dynamism,
(4) Cross-functional spillover emerges mutation which can be enabled by spillover dynamism initiated by interaction of researchers,

(5) Cumulative learning by interactions would be the key in incorporating the new functionality as it constructs the following virtuous cycle:

Cumulative learning → *Functionality development* → *Increase in MPT* → *Increase in TFP growth* → *Increase in GDP growth* → *Increase in learning opportunity* → *Further learning efforts* → *Sustainable coevolution*,

(6) Confronting a post-information society (ubiquitous society), swell of Japan's institutional MOT adapting to new requirements in a ubiquitous society leads high-technology firms to transformation of technopreneurship resulting in the bi-polarization trajectory,

(7) OIR (operating income to R&D) substitution for R&D intensity by means of effective integration of innovation initiative and external acquisition efforts through cumulative market learning would be the key strategy corresponding to such a bi-polarization, and

(8) Institutional shift from co-existence or co-adaptation of homogeneous firms to coevolution among heterogeneous firms would be essential.

11.8 Technological Diversification: Strategic Trajectory Leading to an Effective Utilization of Potential Resources in Innovation: *A Case of Canon*

Prompted by conspicuous Canon's behavior with respect to technological diversification as well as high level of operating income to sales, this chapter attempted to elucidate its dynamism. On the basis of a theoretical relationship between assimilation capacity of spillover technology and diversification of R&D activities, new methodology to measure the extent of technological diversification of firms by means of assimilated spillover technology was developed.

Utilizing this methodology, technological diversification trend in Japan's leading electrical machinery firms was measured and its impacts on R&D intensity was analyzed. In addition, on the basis of a logistic growth function within a dynamic carrying capacity incorporating technology stock, development trajectories of these firms were estimated, thereby marginal productivity of technology, functionality development, TFP growth rate, IRR to R&D investment and impacts of technological diversification on them were analyzed. All correlation analyses among them revealed that Canon demonstrated the highest significance from the late 1980s.

Inducing factors of technological diversification were analyzed and both IRR to R&D investment and sales were identified as sources of Canon's technological diversification.

Noteworthy is that while Canon's technological diversification increased as its sales increased, reverse trends were observed in other electrical machinery firms, which supports a postulate of diversification paradox. This demonstrates that Canon, against such a paradox, sustained its technological diversification consistently by utilizing its sales increase fully.

These results demonstrate that Canon constructed a sophisticated virtuous cycle among high level of IRR to R&D investment, technological diversification and increase in operating income to sales. These observations provide noteworthy suggestions to firms, particularly electrical machinery firms amidst megacompetition as follows:

(1) Canon's technological diversification strategy is beyond the discussion on selection and concentration as well as diversification paradox leading to a new diversification theory in an information society.
(2) Canon's diversification strategy provides a constructive suggestion for effective utilization of spillover technology which is considered the key of competitiveness in an increasing trend in global technology spillover.
(3) Thus, Canon's technological diversification trajectory provides a constructive suggestion to the best utilization of potential resources in innovation by constructing an effective virtuous cycle.

Future works should focus on the adaptability of Canon's business model based on such technological diversification strategy to other firms facing similar situation.

11.9 Japan's Co-evolutionary Dynamism Between Innovation and Institutional Systems: *Hybrid Management Fusing East and West*

On the basis of the foregoing analyses, Japan's coevolutionary dynamism between innovation and institutional systems triggered by hybrid management between indigenous strength (East) and the effects of cumulative learning in 1990s (West) were analyzed. Noteworthy findings and policy implications can be summarized by the following six points:

(1) Coevolutionary dynamism between innovation and institutional systems is decisive for an innovation-driven economy which may stagnate if institutional systems cannot adapt to innovations, and Japan's economy in the 1990s is one example.
(2) Japan indigenously incorporates an explicit function which induces coevolutionary dynamism enabling it to achieve conspicuous performance in a virtuous cycle between innovation and rapid economic growth in the 1960s, followed by technology substitution for energy in the 1970s leading to the world's highest energy efficiency improvement and huge advances in manufacturing technology level in the 1980s.
(3) Although Japan's dynamism shifted to the opposite in the 1990s, resulting in a lost decade because of a systems conflict between indigenous institutional systems and a new paradigm in an information society, a swell of reactivation emerged in the early 2000s.

(4) This can largely be attributed to hybrid management fusing the "East" (indigenous strength) and the "West" (lessons from an IT driven new economy) typically observed in mobile-driven innovation.
(5) Noteworthy success in such hybrid management can be seen in Canon, which effectively utilizes its indigenous strength ("East") in assimilating external technology ("West") while preserving its own organization by not depending on M&A. M&A reacted to deteriorate indigenous organization which does not necessary adapt to exotic systems in Japan. However, as a consequence of the fusing efforts, certain firms have shown noteworthy accomplishments in the synergy of M&A leading to a dramatic increase from 2004.
(6) The effect of fusing "East" and "West" validates the significance of global coevolution that corresponds to SIMOT's aim to elucidate, conceptualize and operationalize Japan's explicit co-evolutionary dynamism. This, then, enables them to accrue as global assets, thereon establishing a new innovative science, the "Science of Institutional MOT," that will enable any country with different institutional systems to effectively utilize its MOT.

The International Institute for Applied Systems Analysis

is an interdisciplinary, nongovernmental research institution founded in 1972 by leading scientific organizations in 12 countries. Situated near Vienna, in the center of Europe, IIASA has been producing valuable scientific research on economic, technological, and environmental issues for over three decades.

IIASA was one of the first international institutes to systematically study global issues of environment, technology, and development. According to IIASA's Governing Council, the Institute's goal is: *to conduct international and interdisciplinary scientific studies to provide timely and relevant information and options, addressing critical issues of global environmental, economic, and social change, for the benefit of the public, the scientific community, and national and international institutions.* Research is organized around three central themes:

- Energy and Technology
- Environment and Natural Resources
- Population and Society

IIASA is funded and supported by scientific institutions and organizations in the following countries:

Austria, China, Egypt, Estonia, Finland, Germany, India, Japan, Republic of Korea, Netherlands, Norway, Pakistan, Poland, Russian Federation, South Africa, Sweden, Ukraine, United States of America.

Further information: http://www.iiasa.ac.at

Printed in the United States
141518LV00005B/25/P

9 783540 892717